CHEMISTRY IN YOUR ENVIRONMENT
user-friendly, simplified science

Every aspect of the world today
- even politics and international relations
- is affected by chemistry

Linus Pauling
Winner of the 1954 Nobel Prize for Chemistry
and the 1962 Nobel Prize for Peace

We chemists have not yet discovered how to make gold
but, in contentment and satisfaction with our lot,
we are the richest people on Earth

Lord George Porter OM FRS
Winner of the 1967 Nobel Prize for Chemistry

CHEMISTRY IN YOUR ENVIRONMENT

user-friendly, simplified science

JACK BARRETT B.Sc., Ph.D., CChem., MRSC
Department of Chemistry
Imperial College of Science, Technology and Medicine
University of London

Albion Publishing
Chichester

First published in 1994 by
ALBION PUBLISHING LIMITED
International Publishers, Coll House, Westergate Street, Aldingbourne, Chichester, West Sussex, PO20 6QL, England
USA: **PAUL & COMPANY PUBLISHERS' CONSORTIUM INC**., P.O. Box 442, Concord, MA 01742

British Library Cataloguing in Publication Data

Barrett, Jack
Chemistry in Your Environment
I. Title
540

ISBN 1-898563-01-2 Library Hardback Edition

ISBN 1-898563-03-9 Student Paperback Edition

Printed in Great Britain by Hartnolls, Bodmin, Cornwall

Table of contents

Preface

This book is a discussion of the impact of the science of chemistry upon our quality of life, as it is experienced in our current environment. It is hoped that readers will be persuaded to agree with the late Linus Pauling's statement quoted on the first page of the book and that they will be persuaded that the influence of chemistry upon the World has been largely beneficial.

The general theme of the book is that the study and application of chemistry has allowed the development and control of our environment to its present level, and will continue to allow that development providing that the Earth's resources are used in a prudent manner.

The book is written principally for readers of the science columns of serious newspapers, and periodicals such as *New Scientist, Science* and *Scientific American.* It will also serve as an introduction to chemistry for readers whose education in that subject was terminated prematurely. It could also serve as an introductory text for students of environmental science. It is hoped that the book will encourage the study of chemistry. Such studies can be very enjoyable and intellectually rewarding. The quotation on the first page of the book, from Lord George Porter's speech at the Brussels symposium to celebrate the 75th anniversary of the founding of the International Union of Pure and Applied Chemistry, is an apt and accurate expression of rewards of a life spent in the study of chemistry.

Chapter 1 is an introduction which indicates the need for chemical knowledge in the understanding of our environment. The fundamentals of chemistry are divided between Chapters 2 and 11. Chapter 2 contains sufficient detail to enable the reader to gain a broad appreciation of the descriptive material in Chapters 3-10. A deeper appreciation of the contents of Chapters 2-10 will be gained by readers who attempt the additional background chemistry explained in Chapter 11. A section which deals with the representations of large and small numbers is contained in the Appendix and there is an extensive glossarial index containing definitions and explanations of technical terms used.

Although some aspects of pollution and its control are mentioned, the subject is not treated at length. Pollution is the responsibility of the human race, chemists and chemistry contributing greatly to its production and control. Pollution may be local or global, short or long term, but we all share the responsibility of making life on Earth as tolerable, and as sustainable, as possible. We can only do this if we ensure that good hard science is applied properly to all aspects of our environment.

It has not been possible in a book of this size to mention all the significant contributions to the subject of chemistry, nor has it been possible to deal with all the effects of chemistry upon our lives.

At the time of writing the U.S.S.R. was undergoing considerable political changes. Because of the time-lag taken to assemble World data, it was inevitable that some data in the book were appropriate to the U.S.S.R. Available data for the C.I.S. have been incorporated where possible.

Acknowledgements

The author is grateful to Eileen Barrett, Ellis Horwood, Kate and Haydn Jones, Leo and Jacqueline McDowell, Alan Otaki, Richard Scorer, Nicole Uprichard, Christine and Dan Van der Vat, and an anonymous referee, for reading parts or all of the manuscript and for making constructive suggestions, all of which have had a significant effect upon the form and content of the final version.

The photograph of the Bingham Canyon mine, used to illustrate the book cover, was kindly provided by the Kennecott Corporation. The author thanks the Corporation for the provision of the photograph and for granting permission for its use in the book.

The majority of the data presented and discussed in the book originated in the following publications:

CRC Handbook of Chemistry and Physics, 73rd. Edition, Ed. David R. Lide, CRC Press, 1992.

Minerals Handbook, 1992-93, by Phillip Crowson, Macmillan Ltd., 1992.

Mining Annual Review, 1991, 1992, published by the Mining Journal Ltd.

The book was created in camera-ready form on a Viglen™ 386/486 computer linked to a Hewlett-Packard™ LaserJet IIP printer. ChiWriter™, version 4.2B, was used for the main text, the periodic tables and some of the chemical structures. Lotus Ami Pro™ for Windows, version 3.0, was used for the title pages and the indexes. The majority of the Figures were produced by Harvard Graphics™, version 3.0, and Harvard Graphics™ for Windows, version 2.0. ISIS™ Draw for Windows, version 1.2, was used for some of the chemical structures.

Kingston upon Thames
September, 1994

Jack Barrett

1

Chemistry in everyday life

Introductory survey of substances and materials which are common in everyday life. The necessity of some understanding and appreciation of chemistry.

Des Carter and his wife had finished their evening meal. The dishes were being thoroughly cleaned by the machine in the new extension to the kitchen. To supplement his intake of fluid, Des was using black coffee followed by a modest volume of a Scottish liquid. Neither of the fluids had any further additives. Des's wife preferred to drink tea with added milk. The couple settled down in their living room to up-date themselves on the World news. They obtained the information from Sky News, a satellite programme which was conducted to their TV set by a copper cable running under the pavement. During the advertising breaks, Des either read the newspaper to check up on the previous days news or attempted to solve the crossword puzzle. He had found that the adverts, particularly those connected with cleaning materials, brought on feelings of nausea and he preferred to avoid them. Even so, the dishes in the machine were being cleaned by one of the advertised products.

It was Winter and the outside temperature was about 3.9°C, thus making it necessary for the Carter's gas fired central heating system to be reinforced by the output from a stove in the living room. The stove was capable of burning anything which was combustible, but the Carters chose to use large nuts of anthracite which were supplemented by bits of their previous fence, and wood which had resulted from the Autumn pruning of the fig tree in the garden.

The Carters were quite conventional and very close to being an average British family consisting of man, wife and two children. To be exactly average they would have had to have produced another 0.2 of a child but this was beyond them. The average family of two parents and 2.2 children did not exist. Both their children had survived the State education system. Joe, the Carter's son, had taken a degree in chemical engineering and was employed by a company as a control engineer. Essentially, his work consisted of the use of computers and the writing of software to control chemical processes. Kate, their daughter, completed her 'A' levels at the local high school and worked initially in a laboratory which controlled the standards of dairy products. She then became a civil servant with the Ministry of Agriculture, Food and Fisheries (MAFF) where she worked in the badger department. Her third job was concerned with corporate pensions and she took postal courses to become professionally qualified. During that time she married a civil servant from MAFF and the two partners carried out a reaction which consisted of mixing their DNA together. The product of this reaction resulted in the birth of a son who was named Ryan. The parents were overjoyed by the event, Ryan's DNA carrying half of each of their own genes. Des and his wife derived the utmost satisfaction from the emergence of the reaction product. After all, the product also contained a quarter of each of their genes.

After their daily dose of information from Sky News and the newspapers, Des and his wife cleared up the kitchen. The waste bins were to be emptied the next morning. Not all the waste was put out for collection. The newspapers, together with a load of unsolicited and unwanted mail, were added to the growing pile of waste paper. Empty bottles, including the one which had contained the wine representing the liquid portion of the evening meal, were placed in a cardboard box. Aluminium cans, mainly those which had contained a dilute coloured solution of phosphoric acid with a sugar-substitute sweetener, were collected in a plastic bag from one of the local supermarkets. The waste paper, glass and aluminium, which were not entrusted to the weekly collectors of rubbish, were delivered at regular intervals to the municipal re-cycling centre, an establishment which used to be called the dump.

The Carters were teachers and writers by profession. They had to do a great deal of thinking whilst preparing for, and carrying out, their business. Des was a very keen thinker, so much so that it prevented his doing as much work as he thought he should! His thoughts were affected by those of a French thinking man. Des was well aware that thinking was a very good indication that he actually existed. He was enthusiastic about continuing to exist and it was this that promoted his thinking activity. There was a suspicion that if he ceased to think, he would cease to be!

After the preparation of the various waste materials, Des took the plastic bags containing the remaining rubbish to the bins in readiness for the next morning's collection. The cold night air had little of the scent of the effluent from the many road vehicle exhaust pipes which caused the day-time rush hour air to smell so badly. The Carter's car was fairly new and contained a catalyst in its exhaust system. The catalyst was supposed to produce a cleaner mixture of exhaust gases. Des drove the car into the garage, closed the garage door and before entering the house looked at the sky. Although the house was in a town, and because there was little cloud cover, Des was able to see the constellation of Orion and, just above the horizon, the bright object of Venus, the evening star. There was a new moon, so that satellite was not visible. Other satellites, however, were obvious and made visible progress on the paths of their orbits. These were the ones which had been placed in the sky by some form of space vehicle. Des remembered the fine sight of a geostationary satellite which had a brightness greater than that of the evening star which he had seen during a visit to Sierra Leone on the west coast of Africa.

Before going to bed that night Des took a drink of water. The water from the tap tasted just a little of chlorine. This gave Des the confidence that the liquid did not contain any viable bacteria. The water Des and Maggie used for the preparation of hot drinks was allowed to pass through a canister in a specially designed jug, whose makers had promised them that the product would be crystal clear and free from any impurities. The use of the purified water in the kettle and coffee machine avoided problems of scaling. They used tap water for that part of their cooking which required the liquid.

Around 3 a.m. Des awoke to emit some liquid waste from his body and, returning to bed, found that his mind was too full of thoughts to allow his return to the sleeping state. His thoughts did not concern his existence or otherwise. He was convinced of his existence. He was much more concerned about why his life was so good. As a sufferer from a poor history curriculum when he was at school, Des became an adherent of Henry Ford. Even so, he was inclined to believe that early men lived in caves and did not have the comforts of modern life. How did this remarkable progress occur? Was it concerned with the exploitation of man's environment? He suspected that it was. His thoughts became concentrated on the details of his surroundings. What did the gases, liquids and solids that surrounded him consist of? Why should the three states of matter, gases, liquids and solids exist? Why were some stable and why did others react together? How far did these reactions go? How fast did the reactions go and what factors determined their rates? How did the reactions occur? He was thinking of questions which are fundamental to the understanding of life as we know it and to the way life is lived. He was not convinced that every component of his surroundings was essential to either his life or to his well-being. He went to sleep.

Des next knew of his existence when the radio came on automatically and he awoke some way through the programme for farmers. There was a discussion of the levels of nitrate (a constituent of some fertilizers, not shift-workers pay) in drinking water. This was followed by a consideration of the drought conditions in part of Europe and there was an attempt to excuse climate change in terms of the greenhouse effect and global warming. The digital thermometer read-out which Des had by his bedside, which recorded the outside temperature, showed a value of 1.4°C and it was raining steadily. His world did not seem to be consistent with what he was hearing. He was well aware that there were variations in the climate. He was originally from Blackburn in the North of England. He knew that the climate in Kingston upon Thames, where he now lived, was appreciably different from that in Blackburn, Kingston being around three hundred kilometres further south than Blackburn. He also knew that on rare occasions the Sun would be shining in Blackburn when rain was falling in Kingston, although normally it was the other way round. He had also come to the conclusion that the recent weather in Kingston was very close to the normal average weather that could be expected in a British winter and that was far from being good! He had a vague memory of summers in Blackburn during his childhood in the 1940s which were far better than any summer that Kingston had had in the last ten years.

The latest, rather jumbled, thoughts of Des were suddenly interrupted by the sound of the door bell. The postman delivered a bulky parcel containing the latest up-grade to Des's word processing program. It had originated in some sunny valley in California, most of its journey being made by air. After a healthy-life breakfast consisting of juice from freshly squeezed oranges and a mixture of prunes from California, sultanas from Australia, oat bran floor-sweepings and milk, Des absorbed the latest news from the papers which had been delivered to the door just after the software had arrived. There was bad news as always. Humans had been exhibiting their almost unique behaviour of maiming and killing members of their own species. In addition to the bad news, there was a report about some Greenpeace personnel trying to prevent a shipment of plutonium leaving from a French port to be delivered to Japan, one of an international conference about atmospheric pollution in Rio de Janeiro and a small item concerning the latest figures about the possible effects of the thinning of the ozone layer. Greenpeace was also advocating that the manufacture of chlorine should cease. It seemed to Des that a considerable degree of chemical knowledge was necessary just to appreciate and understand parts of the daily news.

Coming after his thoughts in the night, these further indications brought Des to the conclusion that he and similarly interested members of the public were in need of a clear and understandable presentation of the scientific principles underlying the effects of man on his environment and the exploitation of the environment for man's benefit. This amounted to a study of the relevant parts of the subjects of physics and chemistry, other science subjects being derived from them. He knew that physics was concerned with the nuclei of atoms and that chemistry was the study of the interactions between atoms and that there was no absolutely clear boundary between the two subjects. He had seen two sets of journals in the library, one called the *Journal of Physical Chemistry*, the other called the *Journal of Chemical Physics*. He thought that a book should be produced which contained an explanation of the science, whatever its name, which governed life and the quality of life.

He discussed the matter with his wife, Maggie, who was an avid reader, accumulator of knowledge and a political libertarian. She was of the opinion that everyone should have the chance of exercising their political will and they should exercise it with as much understanding as possible of the issues of the day. She convinced Des that the main science of which some knowledge would be crucial to the understanding which he sought was chemistry. She suggested that he should contact a chemist who might be interested in writing a book suitable for a wide readership. The book should cover the basics of the science as they affected day-to-day life and the manner in which the science had been applied to produce the added value of the standard of living which Des enjoyed and wondered about. He contacted the present author and, after a few discussions and periods of thought, the book was written. Des insisted that the presentation should be in a logical order with the harder concepts explained more thoroughly for the more ambitious reader. This was so that a reader could, if necessary, take a rapid read through the chapters and pick up a good general appreciation of their content. Des had no difficulties with mathematics but he was aware that the subject did discourage some people. Nevertheless he was of the opinion that any mathematical concepts that were included in the book should be simply, but properly, explained. Another aspect of understanding science was the problem posed by terminology or jargon, those words and symbols which had meaning to the particular scientist but which were practically meaningless to the layman or even a scientist of a different discipline. Chemical equations, which are the chemist's shorthand for representing chemical reactions, would need to be explained in detail. The reader would have to be patient and persistent. The earlier material would be fundamental to the understanding of the subject matter and only after its appreciation would the reader be able to gain full access to its necessary relevance to modern life. Des and Maggie both read the various draughts of the book and made many comments about the content, some of which were incorporated into the final version. Over a celebratory drink of an alcoholic French liquid which emitted bubbles of carbon dioxide they agreed that anyone following their journey through the book would be rewarded by an understanding of life's goodness which would enrich their future thoughts. It might even change their attitude to life and possibly alter their day-to-day activities.

2

Chemistry explained

Elements, atoms and molecules. Compounds. Constituents of atoms. Isotopes. Mixtures and chemical reactions. Chemical equations. Relative atomic mass. The Mole. Valency and chemical combination. Structural formulae and Isomerism. Ions and the electrolysis of water. Acids and bases. Temperature. Chemical and physical changes. Oxidation and reduction.

2.1 INTRODUCTION

This chapter introduces some important principles of chemistry and defines and explains by example some terminology commonly used by chemists. It is hoped that this chapter is sufficient for the reader to be able to appreciate Chapters 3-10 which form the main part of the book. A more extensive survey of the principles of chemistry forms the basis of Chapter 11. The Appendix contains definitions of the units used for quantities of substances, and an explanation of scientific notation for the expression of numbers.

2.2 ELEMENTS, ATOMS AND MOLECULES

A chemical element is a substance that consists of chemically identical atoms, one atom of an element being the smallest particle having the characteristics of that element. There are 109 chemical elements which are classified as belonging to one or other of eighteen groups as is shown by the periodic table in Fig. 2.1. The current total number of elements consists of 88 which are found naturally on Earth, plus 21 which have been synthesized by a variety of nuclear processes. More elements may be synthesized as work on them progresses. The periodic table is so-called because it expresses the periodic law of Mendeleev, first published in 1869, which is that '*the elements, if arranged in order of their atomic weights, exhibit an evident periodicity of properties*'. The modern periodic table is shown in Fig. 2.1 with the elements arranged in order of their atomic numbers. These are related to the electrical charges on their nuclei.

1	2	3	4	5	6	7	8	9	10	11	12	13	14	15	16	17	18
1 H																	2 He
3 Li	4 Be											5 B	6 C	7 N	8 O	9 F	10 Ne
11 Na	12 Mg											13 Al	14 Si	15 P	16 S	17 Cl	18 Ar
19 K	20 Ca	21 Sc	22 Ti	23 V	24 Cr	25 Mn	26 Fe	27 Co	28 Ni	29 Cu	30 Zn	31 Ga	32 Ge	33 As	34 Se	35 Br	36 Kr
37 Rb	38 Sr	39 Y	40 Zr	41 Nb	42 Mo	43 Tc	44 Ru	45 Rh	46 Pd	47 Ag	48 Cd	49 In	50 Sn	51 Sb	52 Te	53 I	54 Xe
55 Cs	56 Ba	71 Lu	72 Hf	73 Ta	74 W	75 Re	76 Os	77 Ir	78 Pt	79 Au	80 Hg	81 Tl	82 Pb	83 Bi	84 Po	85 At	86 Rn
87 Fr	88 Ra	103 Lr	104 Db	105 Jl	106 Rf	107 Bh	108 Hn	109 Mt									

Group number

atomic number

element symbol

57 La	58 Ce	59 Pr	60 Nd	61 Pm	62 Sm	63 Eu	64 Gd	65 Tb	66 Dy	67 Ho	68 Er	69 Tm	70 Yb
89 Ac	90 Th	91 Pa	92 U	93 Np	94 Pu	95 Am	96 Cm	97 Bk	98 Cf	99 Es	100 Fm	101 Md	102 No

Fig. 2.1 A version of the periodic classification of the elements, i.e. the periodic table, showing the arrangements of the elements into eighteen groups; the elements are represented by their symbols; their atomic numbers are shown

There are seven horizontal **periods** of elements which vary in the numbers of elements which they contain. The first contains only the two elements hydrogen (H, Group 1) and helium (He, Group 18). The next two periods, each containing eight elements, (Lithium, Li to Neon, Ne and Sodium, Na to Argon, Ar) are called the first and second short periods respectively. Four long periods follow these with the third and fourth long periods being elongated by having fourteen extra elements each which, to aid the presentation of the diagram, are placed below the main body of the table. The fourth long period is incomplete. The arrangement of the periods allows each of the vertical **Groups** to contain elements which possess similar properties. Detailed descriptions of the periodic table, the names of all the elements and the periodicity of elemental properties are given in Chapter 11.

At the standard temperature of 25°C and at atmospheric pressure some elements are gases, e.g. oxygen and nitrogen; the main constituents of the air, two are liquids, bromine and mercury, but most are solids, e.g. carbon and copper. They exist either as separate atoms or combined either in groups called molecules or in crystals which are infinite arrays of atoms and do not contain groups of atoms identifiable as individual molecules. Each element has a symbol which is derived from its name, or in some cases, the Latin or German equivalents. Helium (symbol: He) is normally a gas consisting of individual helium atoms. Oxygen (symbol: O) is normally gaseous, but consists of molecules which each contain two oxygen atoms. These diatomic molecules are symbolized as O_2 and are called oxygen molecules or more strictly as **dioxygen** molecules. Iron (symbol: Fe from the Latin, *ferrum*) exists as a solid metallic substance in which the atoms of iron are assembled in an orderly crystalline fashion. The assembly has long-range order throughout each crystal of iron and no individual molecules of iron are identifiable. The atoms of iron are arranged in the form of a body centred cubic (bcc) structure. The simplest unit of this structure is shown in the diagram of Fig. 2.2. The architecture of the Atomium museum in Brussels, shown by the photograph in Fig. 2.3, is based upon the bcc arrangement. The 'atoms' contain the exhibits, the links or 'chemical bonds' being connecting walk-ways or escalators. In the structure of the real element the iron atoms are in contact with their nearest neighbours. Diagrams such as that in Fig. 2.2 are drawn deliberately to express clearly the relative positions of the atoms, 'space-filling' models and diagrams being less helpful.

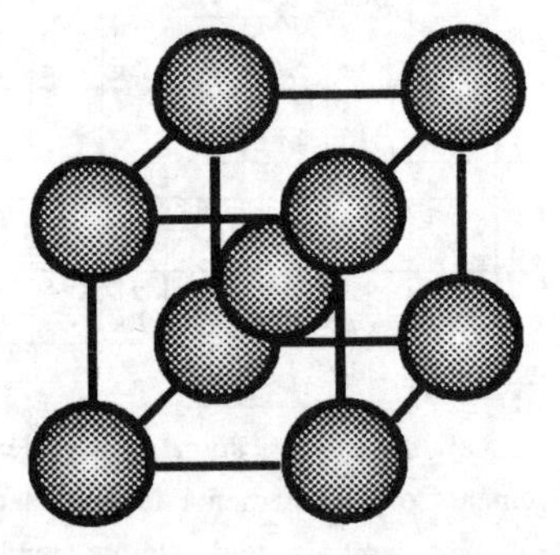

Fig. 2.2 A diagram showing the arrangement of atoms in a body centred cubic structure

2.3 COMPOUNDS

When atoms of different elements combine, they form compounds which may contain molecules with an identifiable formula that indicates the number of each of the contributing atoms, e.g. the formula of water is H_2O, meaning that the water molecule contains two atoms of hydrogen and one atom of oxygen. Other compounds, although having a definite atomic composition, e.g. iron oxide (Fe_2O_3) which exists naturally as the mineral called hematite, do not consist of identifiable molecules. In such cases the solid compound consists of a large array of atoms with long-range order. Compounds, like elements, exist in one or other of the physical states, i.e. gas, liquid or solid, at normal temperatures.

2.4 CONSTITUENTS OF ATOMS, ISOTOPES OF ELEMENTS

An atom consists of a positively charged nucleus surrounded by a sufficient number of electrons to render the atom neutral, electrons being negatively charged. The term 'electron' arises from the Greek word for amber, the fossilized resin which becomes electrically charged when rubbed with a cloth. It should not be thought of as one of Ronald Reagan's slogans! Atomic nuclei contain numbers of positively charged particles called protons, and neutral particles called neutrons. Two important characteristics of an element are its atomic number, Z, which is equal to the number of positive charges on the nucleus, and its relative atomic mass, RAM. The relative atomic mass of an element, which used to be known as its atomic weight, is explained later in this chapter and dealt with in detail in Chapter 11. The structure of an atom, also dealt with in more detail in Chapter 11, may be explained in terms of a positively charged nucleus which is surrounded by a sufficient number of negatively charged electrons to make the atom neutral, i.e. Z electrons. The nucleus consists of a number of protons, each with a positive charge of +1, represented by the value of the atomic number, Z, and a number of neutrons (neutral particles) equal to the value of $A - Z$, where A represents the **mass number**. A is the nearest whole number to the exact relative atomic mass of the atom. In order to describe a particular element in an accurate and concise manner the mass number, A, is quoted as a left-hand superscript to the element's symbol, the atomic number, Z, being indicated as a left-hand subscript:

$$^{\text{mass number}}_{\text{atomic number}}\text{Element symbol; e.g. } ^{35}_{17}\text{Cl and } ^{37}_{17}\text{Cl.}$$

Fig. 2.3 A photograph of the Atomium museum in Brussels, a structure based upon the body-centred cubic arrangement of iron atoms in the crystalline metal [Photo: J.Barrett]

The two different chlorine atoms, so described, are examples of **isotopes** of an element, natural isotopy being a property of most elements. The existence of neutrons made possible the understanding of the existence of isotopes of an element. Isotopes of an element are different atoms which have the same value of Z, otherwise they would be different elements, but have different numbers of neutrons so that they have different values of A. The nuclei of the two isotopes of chlorine both contain 17 protons, the mass 35 isotope containing 35 - 17 = 18 neutrons, and the heavier isotope (mass 37) containing 20 neutrons. Isotopy is discussed more fully in Chapter 11. The atom is rendered neutral by Z negatively charged electrons which occupy the volume of space around the nucleus, e.g. 17 electrons for each chlorine atom.

2.5 MIXTURES AND CHEMICAL REACTIONS

When two elements are mixed together, they might react to form a compound or remain as a **mixture** in which both elements are chemically unchanged. A mixture of iron filings and elemental sulfur (symbol: S) can be made at room temperature in which no chemical reaction occurs. Although the elements are mixed together, they do not react with each other. They could be separated by using a magnet, because iron is magnetic and would adhere to the magnet, or by using carbon disulfide as a solvent for the sulfur, iron being insoluble. If the mixture is placed in a test-tube and heated in a flame, a temperature is reached which causes a violent reaction to occur. Even if the test-tube is withdrawn from the source of heat the reaction continues with the evolution of a great amount of heat. A chemical reaction occurs, the product being iron sulfide which is non-magnetic, and no part of it dissolves in carbon disulfide. The reaction may be described in terms of a chemical equation. The equation which expresses the chemical change and the evolution of heat is:

$$\mathrm{Fe} + \mathrm{S} \longrightarrow \mathrm{FeS} + \text{heat}$$

the product of the reaction, iron sulfide, being written as the formula FeS. The arrow indicates the progression from the reactants (Fe + S) to the product (FeS). Iron sulfide is a solid substance with long-range order, the iron and sulfur atoms forming a particular crystal structure. There are no individual iron sulfide molecules identifiable. That heat is given out in the process is another indication that chemical combination has occurred. The study of heat changes accompanying chemical changes is part of the subject of thermodynamics, i.e. heat movement, which is dealt with in Chapter 11. Production of heat is not associated with all chemical reactions. There are some chemical changes which occur with the absorption of heat.

2.6 CHEMICAL EQUATIONS

The main constituent of natural gas is the hydrocarbon compound, methane, which is a compound containing hydrogen (symbol: H) and carbon (symbol: C) atoms, as the term hydrocarbon suggests, each molecule being composed of four atoms of hydrogen to one atom of carbon. The formulae of other hydrocarbons are dealt with in Chapters 7 and

11. Methane consists of individual molecules which are symbolized as CH_4 to express their atomic composition. When used as a fuel, methane burns in air by reacting chemically with dioxygen molecules to produce carbon dioxide (CO_2) and water (H_2O) molecules. Both carbon dioxide and water exist as triatomic molecules, the subscripts indicating the numbers of atoms of an element in each molecule. If a molecule possesses only one atom of an element no subscript is included in the formula for that element. Each methane molecule produces one of carbon dioxide and two of water. This is because both methane and carbon dioxide molecules contain a single carbon atom, but two water molecules are needed to accommodate the four hydrogen atoms possessed originally by the methane molecule. Such a chemical change may be written symbolically as:

$$CH_4 \longrightarrow CO_2 + 2H_2O$$

the '2' preceding the formula for the water molecule indicating that two water molecules are produced for every methane molecule undergoing the reaction. The '2' is a multiplier for the whole molecular formula it precedes, i.e. $2H_2O = 2 \times (H_2O)$. Normally the parentheses are omitted in chemical equations. The above symbolism does not represent a properly balanced chemical equation. Such an equation must have equal numbers of each kind of atom on both sides of the arrow. Otherwise the indication might be that atoms have either 'come from nowhere' or have 'disappeared', neither of these occurrences being possible. The above incomplete 'equation' may be transformed into a properly balanced equation by counting the oxygen atoms in the products; i.e. two from the CO_2 and two from the two molecules of water making a total of four. Since elemental oxygen exists as dioxygen, two such molecules are required on the left-hand side of the arrow to balance the equation which is then written as:

$$CH_4 + 2O_2 \longrightarrow CO_2 + 2H_2O$$

Chemical equations express accurately the composition of the reactants and the products and the ratio of the numbers of molecules which are necessary to allow the reaction to occur. In this example it is necessary for there to be two molecules of dioxygen for every molecule of methane burned, and the equation also indicates that every molecule of methane is transformed into one molecule of carbon dioxide and two molecules of water. Chemical equations are not just a shorthand method of describing reactions. They give precise information about the amounts of the participating reactants which are needed to produce certain amounts of the products.

The extraction of iron from its oxide mineral, hematite, is an example of the importance of chemical equations. Hematite is a non-molecular solid substance with the formula Fe_2O_3. The oxygen content is removed by reacting the oxide with carbon monoxide gas, consisting of CO molecules, in a blast furnace, the equation being:

$$Fe_2O_3 + 3CO \longrightarrow 2Fe + 3CO_2$$

The equation indicates that three molecules of carbon monoxide are needed to react with the three oxygen atoms in the hematite 'formula unit' to allow the production of elemental iron. The carbon monoxide used in the process is produced by burning carbon (in the form of coke) in oxygen, the equation for this being:

$$2C + O_2 \longrightarrow 2CO$$

The problem confronting the iron manufacturer is to know how much coke to mix with the hematite in order to extract all its iron content. The manufacturer would know by chemical analysis the hematite content of the feed material which was to have its iron content extracted. The furnace might hold a charge of 100 tonnes of hematite. The tonne is the unit of weight (strictly mass) sometimes known as the 'metric ton', equal to 1000 kilograms or a million grams. The problem is to find the mass of coke which will provide sufficient carbon monoxide to convert the 100 tonnes of hematite into iron and to find the mass of iron produced. Before the calculation can be carried out some knowledge of atomic masses is essential.

2.7 RELATIVE ATOMIC MASS (RAM)

The basis of atomic masses is dealt with in detail in Chapter 11 and is only outlined in this section. Atoms are very small particles with extremely small masses which are not usually used in chemical calculations. The quantities which are normally used are relative atomic masses (RAMs) which used to be called atomic weights. The RAM values needed for the calculation started in Section 2.6, and for subsequent examples, are given in Table 2.1.

Table 2.1 - Relative atomic masses of elements discussed in this chapter

element	RAM
hydrogen	1
carbon	12
oxygen	16
sulfur	32
chlorine	35.5
potassium	39
iron	56

The values are quoted as whole numbers, except that of chlorine, to simplify the arithmetic.

The calculation of the expected yield of iron metal from 100 tonnes of hematite is carried out in the following manner. The relative mass of a formula unit of hematite (Fe_2O_3) is calculated as:

$$(2 \times 56) + (3 \times 16) = 160$$

The relative mass of the contained iron is 2 x 56 = 112. This indicates that every 160 units of hematite contains 112 units of iron, whatever the units are. So 100 tonnes of hematite should contain $100 \times \frac{112}{160} = 70$ tonnes of iron. The manufacturer knows to expect that quantity of iron to be released from the furnace when the reaction is complete.

The calculation of the amount of coke to be used is best carried out in two stages, the first being the calculation of the amount of carbon monoxide required. The relative atomic masses of carbon and oxygen are 12 and 16 respectively, so that their sum, 28, represents what is known as the **relative molecular mass** (RMM) of carbon monoxide. It is not possible to assign such a description to a non-molecular solid such as hematite, but the use of the relative formula mass of 160, as calculated above, is adequate and permissible. The equation for the reaction of hematite with carbon monoxide indicates that three relative mass units of carbon monoxide (3CO) are needed for each relative mass unit of hematite. This means that for the reaction of 100 tonnes of hematite, the mass of carbon monoxide needed is:

$$\frac{100 \times (12 + 16) \times 3}{160} = \frac{100 \times 28 \times 3}{160} = 52.5 \text{ tonnes.}$$

The second stage of the calculation is to find the mass of carbon required to produce 52.5 tonnes of carbon monoxide. Twelve (RAM of carbon) units of carbon will be contained by 28 (RMM of CO) units of carbon monoxide. 52.5 tonnes of carbon monoxide contain $52.5 \times \frac{12}{28} = 22.5$ tonnes of carbon.

The reader will appreciate that calculations of this kind are of vital importance in chemical industry. They indicate the relative amounts of reactants which are necessary for any chemical process and are an essential part of the calculations upon which the economics of any process depend. If the manufacturer uses only 20 tonnes of coke for the 100 tonnes of hematite some, of the oxide will be unaffected by the extraction process. If the manufacturer were to use 25 tonnes of coke the iron business would lose money, but the coke manufacturer would do well until the iron manufacturer went out of business!

2.8 THE MOLE

The contributions to chemical theory by Gay-Lussac and by Avogadro are discussed in further detail in Chapter 11, but some mention of them in this chapter is essential for the further understanding of chemical equations. Gay-Lussac studied reactions that occurred between gases and gave gaseous products. His law states that *'there is a simple relationship between the volumes of the reactants and those of the products'*. This is demonstrated by the gaseous reaction of dihydrogen with dioxygen to give water which may be written as the equation:

$$2H_2(g) + O_2(g) \longrightarrow 2H_2O(g)$$

which is chemists' shorthand for recording that two molecules of gaseous diatomic hydrogen (dihydrogen, the 'g' in parenthesis indicating the gaseous state of the substance) react with one molecule of gaseous diatomic oxygen (dioxygen) to give two molecules of gaseous water. In accordance with the Gay-Lussac law, two volumes of dihydrogen react with one volume of dioxygen to give two volumes of gaseous water under the conditions of the reaction i.e. at a sufficiently high temperature to ensure that the water is in the gaseous state.

Avogadro's highly important contribution to the understanding of the molecular nature of elements and compounds was to explain the connexion between the numbers of reacting **molecules** and their reacting **volumes**, and is expressed in terms of his famous hypothesis. This is crucially dependent upon the difference between atoms and molecules, e.g. that hydrogen gas consists of dihydrogen molecules, H_2, and is not an assembly of hydrogen atoms, H. The Avogadro hypothesis is *'that equal volumes of gases, at the same temperature and pressure, contain equal numbers of molecules'*. Thus, equal volumes of dihydrogen and dioxygen contain the same number of molecules and, in order to form water, the volume of dihydrogen must be twice that of dioxygen.

The reacting ratio of two molecules of dihydrogen to one molecule of dioxygen is the consequence of the formula of the water molecule and the diatomic nature of the reactant molecules. If elemental hydrogen and oxygen were both monatomic, the equation for their reaction to produce water would be:

$$2H(g) + O(g) \longrightarrow H_2O(g)$$

and would imply, via Gay-Lussac's law, that two volumes of hydrogen would react with one volume of oxygen to give only **one** volume of water.

Chemical equations can also be interpreted in terms of the masses of the particpating substances. Taking the relative atomic masses of hydrogen and oxygen as 1 and 16 respectively, it is possible to interpret the reaction between dihydrogen and dioxygen in terms of relative molecular mass. For example, taking the gram as the unit of mass, four grams of dihydrogen ($2H_2$ = 4H; 4 x RAM(H) = 4) react with thirty two grams of dioxygen (O_2 = 2O, 2 x RAM(O) = 32) to give thirty six grams of water (2 x RMM(H_2O) = 36).

The connexion between relative molecular masses and the number of molecules is that the number of molecules in four grams of gaseous dihydrogen must be twice the number of molecules in thirty two grams of gaseous dioxygen.

The above relationships are summarized in Table 2.2.

The mass of a molecule represented by the relative molecular mass (RMM) expressed in grams is an important quantity and is known as one **mole**, not to be confused with the furry animal. Thus, one mole of dihydrogen consists of two grams of the gas, one mole of dioxygen is 32 grams and one mole of water is 18 grams.

A consequence of the above relationships is that, under the same conditions of temperature and pressure, one volume of any gaseous element or compound contains the same number of molecules. The gram molecular volume, which is the volume occupied by one mole of a gaseous substance, is now generally accepted to be 24.465 litres at a temperature of 25°C and at standard atmospheric pressure which is 1013.25 millibars. The number of molecules contained by one mole of a substance, whatever its physical state, i.e. gas, liquid or solid, is known as Avogadro's number or Avogadro's constant. The number is 6.022 x 10^{23}. The constant has a value of 6.022 x 10^{23} per mole, which applies to the number of particles in one mole of the substance, i.e. the number of molecules of dihydrogen in 2 grams of the element, the number of molecules of dioxygen in 32 grams of the element. The number is normally expressed in the scientific notation above (see the Appendix for an explanation of scientific notation), but if written out fully is 602200,000000,000000,000000. In words it would be six hundred and two thousand two hundred, million million million.

Table 2.2 - The reaction between dihydrogen and dioxygen in terms of numbers of molecules, volumes of gas, masses and relative molecular masses of participating elements and compounds

molecular units	$2H_2(g)$	$O_2(g)$	$2H_2O(g)$
number of molecules	2	1	2
volumes of gas (e.g. L)	2	1	2
masses/grams	4	32	36
relative molecular mass	2 x 2 = 4	32	2 x 18 = 36

The extension of the mole concept to non-gaseous substances is of considerable importance in the interpretation of chemical equations. In the reaction of iron with sulfur, the chemical equation may be interpreted in terms of one mole of iron (56 grams) reacting with one mole of sulfur (32 grams) to give 56 + 32 = 88 grams of iron sulfide as the product.

The mole concept allows chemists to have the confidence that they can construct molecules, within the allowable combining powers of the elements, from ingredients whose masses are in a suitable ratio dependent upon their relative atomic and molecular masses. Chemistry is normally carried out in terms of molar quantities, or appropriate multiples or fractions of moles, which avoids the counting out of appropriate numbers of atoms and molecules; that being practically impossible. Many chemical reactions are carried out in solutions in which the mole concept is very useful in defining the **concentrations** of the reactants. A solution is made by dissolving an amount of a compound (called the solute, usually a solid) in a liquid (called the solvent). The unit of volume is the litre (L) which is sub-divided into one thousand millilitres (mL) or cubic centimetres (cc). The litre is the volume of a cube with sides which are ten centimetres (10 cm = one decimetre, 1 dm) long. One litre is a cubic decimetre. If one mole of a substance is dissolved in one litre of solution, the concentration of that substance is one molar, which may be expressed as 1 M or more strictly as 1 mole per cubic decimetre. A 1 M aqueous solution of potassium chloride, KCl, would be made by dissolving 74.5 grams of the substance in water so that the final volume of the solution is one litre. The 74.5 is the relative molecular mass or relative molar mass (RMM) of potassium chloride, made up from contributions of 39 and 35.5 which are the relative atomic masses (RAMs) of potassium and chlorine respectively (see Table 2.1).

2.9 VALENCY AND CHEMICAL COMBINATION

Atoms which enter into chemical combination have a combining power which is known as their **valency**. The word valency is derived from the Latin word meaning 'to be worth'. When atoms combine, they are considered to form **chemical bonds.**

The subject of valency is treated in some depth in Chapter 11, but some appreciation of it is necessary before reading Chapters 3-10. A few examples of compounds, already mentioned in this chapter, suffice to introduce the subject. There are three main ways in which atoms combine with each other. One is for two atoms to share two electrons in forming what is known as a **covalent bond.** The two electrons, being negatively charged, act as a 'glue' holding together the remainders of the two otherwise neutral atoms, which may be thought of as two positively charged particles, viz:

$$\oplus \ \overset{-}{\underset{-}{}} \ \oplus$$

The second form of bonding is **ionic**, consisting of the transfer of an electron from one atom, which leaves it positively charged, to another, producing a negatively charged particle. Such charged particles are called ions. The resulting substance is held together by the electrostatic attractions between the oppositely charged ions, M^+X^-.

A third form of bonding consists of two atoms sharing two electrons which are provided by one of the atoms. This is called **co-ordinate** or **dative** bonding.

Hydrogen, in its elemental form, exists as the molecule, H_2, the two electrons being shared between the two nuclei (protons, H^+) in the formation of a single covalent bond. The valency of hydrogen is thus considered to be 1. The molecule of methane, CH_4, shows that the valency, i.e. the combining power, of the carbon atom is four, because it forms bonds with four monovalent hydrogen atoms. In potassium chloride, KCl, the potassium atom loses one electron which is accepted by the chlorine atom. Both elements are monovalent in forming this particular ionic compound. The solid contains equal numbers of positive potassium ions and negative chloride ions. All elements have characteristic valencies, some exhibiting more than one value, three of the group 18 elements - helium, neon and argon - possessing zero valency. The valencies of elements determine the formulae of compounds that they form.

The concept of valency rationalizes the general observation that compounds have constant compositions with fixed simple ratios between the numbers of atoms they contain. Potassium chloride always contains an equal number of potassium and chlorine atoms because both elements are monovalent. The water molecule always has twice as many hydrogen atoms as oxygen atoms because hydrogen is monovalent and oxygen is divalent. The terms monovalent and divalent are synonymous with the terms univalent and bivalent, the prefixes being derived from Greek or Latin words for single and double.

2.10 STRUCTURAL FORMULAE AND ISOMERISM

Although the atomic composition of a molecule may be adequately represented by a written formula, there are many cases where it is not satisfactory. A structural formula represents the exact connectivities of the atoms of which the molecule is constituted. In the simple case of the methane molecule the structural formula could be represented by the diagram in Fig. 2.4, from which it is clear that all four hydrogen atoms are connected, or **bonded**, to the central carbon atom.

This two-dimensional representation is a satisfactory method of indicating connectivity, but fails to show that the four hydrogen atoms are arranged at the four vertices of a regular tetrahedron, the carbon atom being in the centre of the tetrahedron. The diagram of Fig. 2.5 shows two views of the three-dimensional methane molecule. One view is obtained by placing the hydrogen atoms at alternate corners of a cube with the carbon atom at its centre. This might seem an over-elaborate method of drawing a structural formula but the three-dimensional nature of atomic connectivity in more complex molecules is extremely important. The view of tetrahedral methane on the right-hand side of Fig. 2.5 is the more conventional one.

```
    H
    |
H—C—H
    |
    H
```

Fig. 2.4 A two-dimensional representation of the structural formula of methane indicating that all four hydrogen atoms are connected to the central carbon atom

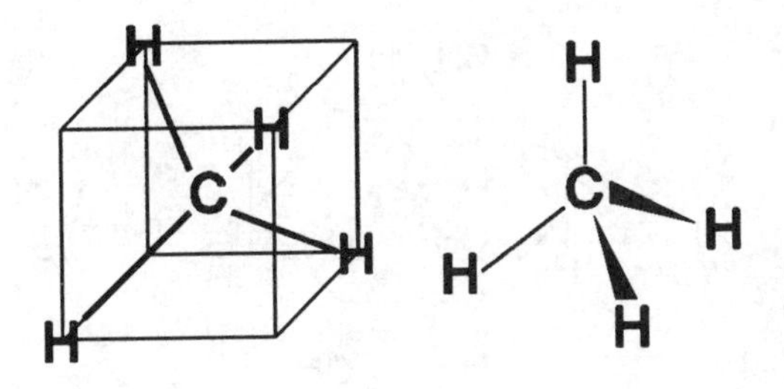

Fig. 2.5 Two views of the tetrahedral methane molecule; the diagram on the left shows the relationship of the tetrahedron to the cube; the diagram on the right is the more common method of signifying the three-dimensional tetrahedral structure of the molecule

The formulae of the compounds, ethanol (or ethyl alcohol, all alcohols contain the C-OH atom grouping indicated by the -ol ending to their names) and dimethyl ether (an ether contains the C-O-C atom grouping), are identical; C_2H_6O. Their molecular structures differ in the connections between the atoms, all connectivities being consistent with the normal valencies of 1, 2 and 4 for the hydrogen, oxygen and carbon atoms respectively. The two-dimensional structural formulae of the two compounds, shown in Fig. 2.6, are sufficient to distinguish one from the other as are the more concise formulae which are also included in the diagram. Such structural representations of molecular structures are meant to indicate the atom connectivities only. They are not used to indicate molecular shapes. The two molecules in Fig. 2.6 are three-dimensional with a tetrahedral arrangement of atoms around each carbon atom, the C-O-H and C-O-C angles being around 120°.

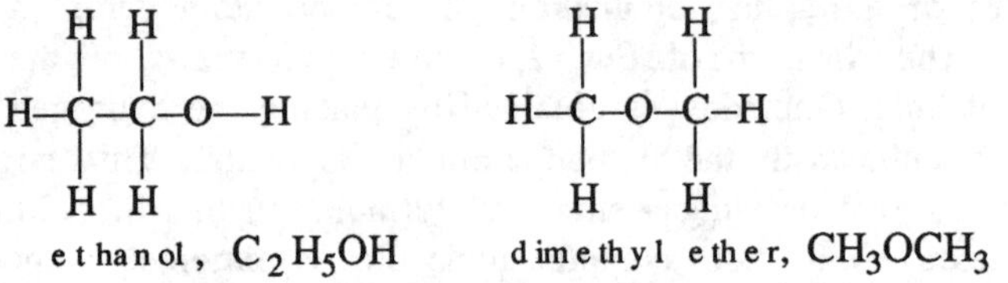

ethanol, C_2H_5OH dimethyl ether, CH_3OCH_3

Fig. 2.6 Structural, and more concise, formulae of the molecules of ethanol and dimethyl ether; the formal name for the latter is methoxymethane.

The two molecules are examples of structural isomerism, they are structural isomers; molecules with identical atomic composition but which have different structures. This example of isomerism shows the importance of the phenomenon, ethanol being a liquid at room temperature, and the active constituent of some nutritive beverages, while dimethyl ether is normally a gas at room temperature and has anaesthetic properties if breathed in. A more complex example of isomerism is given by the structures of the α and β forms of D-glucose, both of molecular formula $C_6H_{12}O_6$, shown in Fig. 2.7.

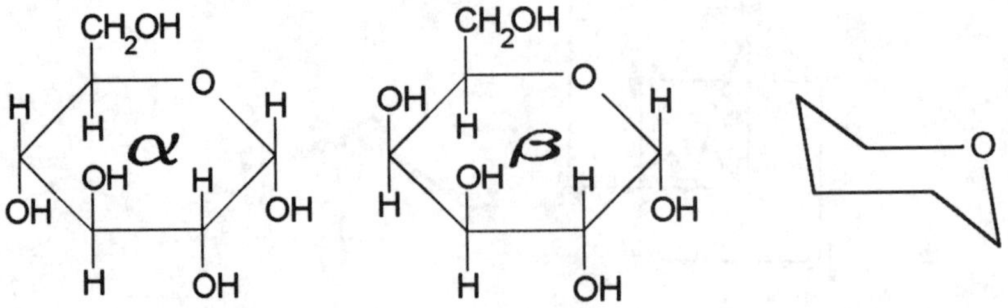

Fig. 2.7 The molecular structures of α- and β-D-glucose and the 'chair' arrangement of the ring atoms; in both of the structures the five carbon atoms which, with the oxygen atom, form the six-membered ring arrangement are not shown by the element symbol; the structures ignore the detailed geometry of the ring which is not planar, but is 'chair' shaped as indicated

In these structures, five of the carbon atoms are not indicated as such. Together with the oxygen atom, they form a six membered 'ring' arrangement and their presence in the molecule must be inferred from the intersection of four 'chemical bonds'; the single lines which are commonly used to demonstrate atom connectivities. This form of shorthand is very frequently used in drawing the structures of organic compounds, i.e. molecules containing carbon. The two isomers of glucose differ only in the positions of the hydrogen atoms and the hydroxyl (OH) groups on the carbon atom on the left-hand sides of the structures shown in Fig. 2.7. The α form is known as D-glucose or dextrose. It is a very sweet tasting sugar which is important in the body's metabolism of foods. The β form cannot be metabolized by the human body.

The structures of organic compounds are conventionally drawn with lines to represent the bonds between the carbon atoms. The other atoms are then shown attached to the carbon skeleton, usually with the omission of any hydrogen atoms which are attached to the carbon atoms. The details of such structures are given in Chapter 11 and are mentioned whenever necessary in the text of the intervening chapters.

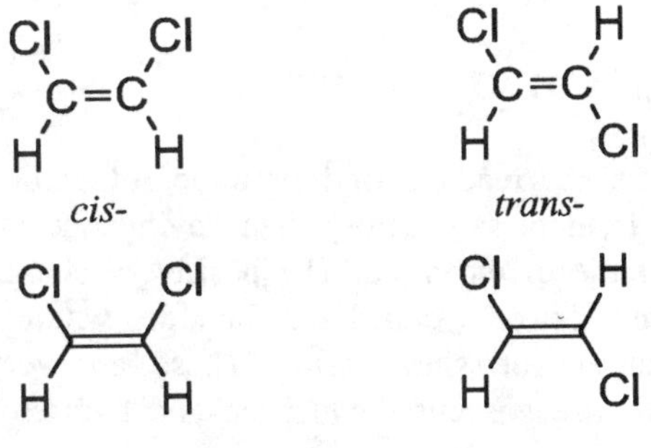

cis- *trans-*

Fig. 2.8 Structural formulae for the two isomers of the 1,2-dichloroethene molecule, showing the *cis-* and *trans-* geometrical arrangements of the two chlorine atoms with respect to their placement at either side of the carbon-carbon double bond. The structures may be drawn with the carbon atoms indicated or with implied carbon atoms as in the lower diagrams

An example of geometrical isomerism is shown in Fig. 2.8 for the *cis-* and *trans-* structures of the compound, 1,2-dichloroethene, $C_2H_2Cl_2$. The nomenclature implies that there is one chlorine atom bonded to each of the carbon atoms. In this compound the carbon-carbon linkage is a double bond consisting of two pairs of shared electrons. The two isomers of the compound are planar molecules, the difference between them being in the relative placements of the hydrogen and chlorine atoms with respect to the C=C bond. In the *cis*-isomer the chlorine atoms are at the same side of the C=C bond, whereas in the *trans*-isomer a chlorine atom is placed at either side of it.

2.11 IONS AND THE ELECTROLYSIS OF WATER

A solution of potassium chloride in water is a good conductor of electricity. Pure water is a very poor conductor of electricity and in order for aqueous solutions to be able to conduct they must contain ions, these being charged particles. Potassium chloride, and the more familiar sodium chloride (common salt, NaCl), are ionic compounds. Even in the solid crystal they contain positively charged and negatively charged ions. When potassium chloride, which consists of equal numbers of positive potassium ions (K^+) and negative chloride ions (Cl^-), is dissolved in water, the solid dissociates into its separate ions which are then free to move independently of each other throughout the volume of the solution. When two electrodes, i.e. wires attached to the two poles of a battery, are placed in the solution an electric current flows. Complex changes occur which may be understood in terms of some possible intermediate processes. The negatively charged chloride ions are attracted to the positive electrode, called the anode, where they give up their electrons to the electrode, thereby becoming neutral chlorine atoms. These are very unstable and react rapidly with water to produce gaseous dioxygen and hydrogen ions (H^+):

$$Cl^- \longrightarrow Cl + e^- \text{(to the anode)}$$

$$4Cl + 2H_2O \longrightarrow 4Cl^- + 4H^+ + O_2$$

It is necessary to consider the reaction of four moles of chlorine atoms to give one mole of dioxygen. This is more satisfactory than having one mole of chlorine atoms produce a quarter of a mole of dioxygen. The positively charged potassium ions are attracted to the negative electrode, called the cathode, where they each acquire an electron to become neutral potassium atoms. These are very unstable and react rapidly with water to produce gaseous dihydrogen and hydroxyl ions (OH^-):

$$K^+ + e^- \text{(from the cathode)} \longrightarrow K$$

$$2K + 2H_2O \longrightarrow 2K^+ + 2OH^- + H_2$$

the second equation being written so that no fractions of moles are required. Any hydrogen ions and hydroxyl ions which are produced react very rapidly with each other to produce neutral water molecules:

$$H^+ + OH^- \longrightarrow H_2O$$

The overall effect of the flow of current is the electrolysis, i.e. the splitting apart by electricity, of water, the final products being dihydrogen and dioxygen in a volume ratio and a molar ratio of two to one (2:1) in accordance with the equation:

$$2H_2O \longrightarrow 2H_2 + O_2$$

This equation is the reverse of that which describes what happens when the two product gases react together in a flame. In the case of the electrolysis, energy from the battery is used to make the reaction occur. When the two gases burn, energy is released in the form of heat.

2.12 ACIDS AND BASES (OR ALKALIS)

Acidity and alkalinity are properties of solutions of substances in water or other solvents. This introductory discussion is restricted to solutions of compounds in water. An acid may be defined as a compound which dissolves in water to produce aquated or hydrated hydrogen ions, $H^+(aq)$, a base or alkali giving hydrated hydroxide ions, $OH^-(aq)$. The extent of the acidity or alkalinity of a solution is usually measured by the value of its pH. The pH value of a solution is used because of the great range of values of the concentration of hydrated protons or hydrogen ions produced in solutions of various acids and bases.

Table 2.3 - Approximate pH values for some common liquids

liquid	pH value
battery acid	1.0
lemon juice	2.5
vinegar	2.8
apple juice	3.0
cola drinks	3.5
tomato juice	4.7
normal rain water	5.5
milk	6.5
pure water (gas free)	7.0
blood	7.4
bicarbonate solution	8.4
milk of magnesia	10.5
ammonia solution	12.0

The prefix, p, which means take the logarithm of whatever value it precedes and then change its sign, arises from the German word *potenz* meaning power, exponent or potency, the H referring to the concentration of the hydrogen ion. The pH scale extends from a value of 0, corresponding to a very acidic solution, to a value of 14 which corresponds to a very alkaline solution. The pH of pure water (regarded as a neutral liquid) is 7.0; acidic solutions have pH values less than 7, alkaline solution have pH values greater than 7. The logarithmic nature of the scale implies that there is a dilution of the hydrogen ion concentration by a factor of ten for each increase of 1 of the value of pH, e.g. a solution with a pH value of 3 is ten times less acidic than a solution with a pH value of 2. The approximate pH values of some common liquids are given in Table 2.3.

The use of litmus paper allows acidic and alkaline solutions to be distinguished. The paper is coloured red in contact with an acid solution (pH < 7) and blue in an alkaline solution (pH > 7). Litmus is an extract from naturally growing lichens, other natural indicators being made by various vegetable extracts such as the water which has been used for cooking red cabbage or beet-root. With **edible** solutions and liquids the decision as to whether they are acidic or alkaline can easily be made by tasting them. **Such methods are very crude and possibly dangerous,** but there is a set of indicators which change colour over different ranges of pH values and which can be used, in paper form, to measure the pH value of any solution to an accuracy of ±0.1 units. For more accurate measurement of the pH value of a solution, a pH meter which has been suitably calibrated with standard solutions is used.

The effect of pH changes on the colour of some molecules is sometimes quite spectacular. The colours of red poppies and blackberries arise from the different selective absorptions of different wavelength ranges of visible light by the same molecule under differing conditions of acidity. A complex compound, with the complicated name, cyanidin-3,5-diglucoside, which may be written as the simple formula RH, which singles out one hydrogen atom, the remainder of the molecule being represented by the group R, is responsible for the deep violet colour of blackberries and black currants. The compound is a violet colour only when the pH value is around 7, i.e. neutral. In acidic solutions a hydrogen ion attaches itself to the molecule to give the ion, RH_2^+, which has a red colour. The same red ion is responsible for the colour of red poppies, roses, rhubarb and red cabbage, chemically similar ions giving the colours of red asters, red apples, copper-beech leaves, cherries and antirrhinums. The saps in these species are acidic. It was thought for many years that the blue colour of the cornflower arose from the selective absorption of light by the ion R^- produced from the molecule RH in alkaline conditions. The molecule does dissociate in alkaline solution to give a blue coloured solution, but no plant sap is alkaline. Cornflower blue is produced by a complex interaction between the above molecule, another similar molecule, and ions of magnesium and iron.

The strongest acid encountered around the home (hopefully, in the garage) is that found in the lead/acid batteries used in cars. It is a concentrated solution of sulfuric acid which has the formula, H_2SO_4. The concentrated acid is known sometimes as oil of vitriol, the pure liquid being quite viscous. The strongest alkali, or base, in the home is sodium hydroxide (NaOH), sometimes known as caustic soda. An important property of acids and bases is that they neutralize each other when mixed together in the correct proportions. A mixture of sulfuric acid and sodium hydroxide undergoes the reaction:

$$H_2SO_4 + 2NaOH \longrightarrow Na_2SO_4 + 2H_2O$$

the products being the salt called sodium sulfate, and water. **The reaction is accompanied by the release of a large amount of heat and should not be attempted.**

Both reactants are dangerous to use and should be removed from the skin by washing with copious amounts of water if contact is made. Sodium bicarbonate is a safe neutralizer of acids, vinegar (its acidity being due to the presence of acetic acid solution, CH_3COOH) is a safe neutralizer of alkalis.

2.13 TEMPERATURE; CHEMICAL AND PHYSICAL CHANGES

It is well known that solids melt and liquids boil when they are heated sufficiently. The understanding of such physical changes; transitions from solid to liquid and from liquid to gas phases, is closely linked to the concept of temperature. The temperature of a gas is proportional to the average speed of its molecules. The detection of the temperature of a gas, such as the air when it is in contact with the human skin, is dependent upon the effects of the molecular collisions on nerve endings. The faster the molecules are, the higher is the perceived temperature. The same is true of liquids, the molecules having similar translational motion, i.e. movement in three dimensions to any part of the occupied volume, as do gaseous molecules. The constituent units of solids have no translational freedom but, in aggregate, they have a very large number of possible vibrations and it is the transfer of the vibrational energy to nerve endings which allows our perception of their temperature. It is common knowledge that hot objects, left to themselves, cool down. The reverse process never happens spontaneously: heat has to be supplied to raise the temperature of a cold body. A mixture of a volume of water at 55°C with the same volume of water at 19°C quickly acquires a temperature of $(55 + 19) \div 2 = 37$°C. Humans experiment in this way every time they fill a washbasin. The combined volume of water assumes an average temperature depending upon the relative volumes of the mix of hot and cold water and on their two temperatures. The mixture of hot and cold water cools if its temperature is higher than that of the room. In scientific reports it is conventional to record temperatures in degrees Celsius (°C), but Charles's law, '*the volume of a gas decreases as the temperature decreases*', indicates that there is an absolute zero, i.e. the temperature at which the volume of the gas is zero, at -273.15°C. The basis for the absolute scale of temperature (the Kelvin scale) is that 0 K = -273.15°C. A temperature on the Celsius scale can be converted into one on the Kelvin scale by the addition of 273.15:

$$K = C + 273.15 \text{ or } C = K - 273.15$$

The Kelvin scale was named after William Thomson who became Lord Kelvin. His early work led to the theory and practice of refrigeration, hence the name Kelvinator for some early domestic and commercial refrigerators. The Fahrenheit scale, not used in scientific reporting, was derived from the 'fixed points' of temperature given by an ice/salt mixture of 0°F, the salting of icy roads causing melting of ice above this temperature, and of the blood of humans, designed to be 100°F, but because of inaccuracies became 98.4°F. Maybe Fahrenheit had a fever at the time of the first calibration. On that scale the melting and boiling points of water are 32°F and

212°F respectively. The 180 degrees Fahrenheit difference is to be contrasted with the 100 degrees Celsius difference, which leads to the conversion formulae:

$$C = (F - 32) \times 5 \div 9$$

$$F = (C \times 9 \div 5) + 32$$

where C and F represent the temperature in degrees Celsius and Fahrenheit respectively.

The differences between the two scales of temperature are shown in Fig. 2.9, together with their relationship to the absolute Kelvin scale. The diagram shows that 100 Celsius degrees are equivalent to 180 Fahrenheit degrees, which is the origin of the 5:9 ratio used in the conversion formulae.

Heat, besides being produced by some chemical reactions, when applied externally to a reaction mixture of substances which react chemically with each other, usually causes the reaction to go faster. This is well known by cooks who use elevated temperatures to make foods more edible and digestible. The application of heat to solid substances can cause them to melt to form liquids. Further heating causes the liquid to boil and become gaseous. Such physical changes, i.e. changes of physical state; solid ⟶ liquid, and liquid ⟶ gas, occur at widely different temperatures for different substances, depending upon their individual structures. The variation ranges from monatomic helium, with a melting point of 3.5 K and a boiling point of 4.3 K, to metallic tungsten, the respective temperatures being 3680 K and 6200 K.

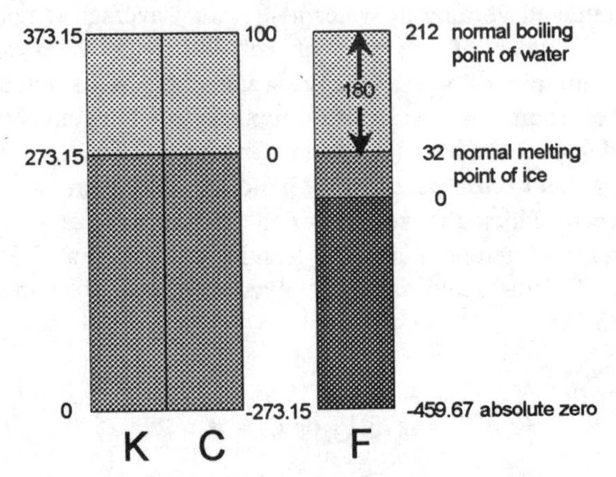

Fig. 2.9 A diagram comparing the Kelvin (K), Celsius (C) and Fahrenheit (F), scales of temperature between absolute zero and the normal boiling point of water, i.e. under standard pressure of one atmosphere at sea level

2.14 OXIDATION AND REDUCTION

The terms oxidation and reduction are used to describe many chemical reactions. In the production of iron from its oxide by reaction with carbon monoxide:

$$Fe_2O_3 + 3CO \longrightarrow 2Fe + 3CO_2$$

the hematite may be said to have been reduced to the element. The carbon monoxide may be described as the reducing agent. The reducing agent becomes oxidized to carbon dioxide; i.e. by the addition of an extra atom of oxygen to the CO. In reactions in which oxidation occurs there must be an equivalent amount of reduction. As the song about *Love and Marriage* goes, '*you can't have one without the other*'. This is now possibly an out-dated concept, but applies precisely to oxidation and reduction.

The oxidation/reduction concept is applied more widely to reactions in which oxygen atoms do not nessessarily participate. In general, oxidation is regarded as the loss of electrons, reduction occurring as electrons are gained. In the combined state, oxygen may be considered to be the oxide ion, O^{2-}, produced when the atom gains two electrons, i.e. is reduced. In the example above, the compound Fe_2O_3 may be regarded as containing three oxide ions with a combined electrical charge of -6. This must be balanced, because the compound is neutral, by the two iron atoms sharing a charge of +6. This would indicate that the compound contained two iron ions, each with a charge of +3, ($2Fe^{3+}$). It would be true if the iron oxide were a completely ionic compound, which is far from being the case, but the concept is useful because it allows some rationalization of a great number of reactions. The ideas are further refined in terms of the allocations of oxidation numbers or **oxidation states** to atoms in compounds. Thus the oxidation state of oxygen is usually -2 and that of iron in hematite is +3. It is conventional to express oxidation states as Roman numerals: thus iron in its +3 oxidation state would be written as Fe^{III}, and oxygen in its -2 oxidation state would be written as O^{-II}. The oxidation state of any element in its elemental form is zero. Applying the ideas to carbon monoxide, the carbon would be considered to be in its +2 oxidation state, i.e. C^{II}, because oxygen(-II) is the other constituent of the molecule. In carbon dioxide the carbon is considered to be in its +4 state, i.e. C^{IV}, since there are two O^{-II}s present in CO_2. The chemical equation may be re-written with the oxidation states of the elements indicated by Roman superscripts:

$$Fe_2^{III}O_3^{-II} + 3C^{II}O^{-II} \longrightarrow 2Fe^{0} + 3C^{IV}O_2^{-II}$$

This equation is helpful in its indication that the oxidation state of oxygen in the three compounds in which it appears in the equation is constant at -II. The iron is reduced from its III state to zero, the carbon being oxidized from II to IV. A simpler equation which expresses the changes in oxidation states is:

$$2Fe^{III} + 3C^{II} \longrightarrow 2Fe^{0} + 3C^{IV}$$

$$\underbrace{\qquad 6e^- \qquad}_{3C^{II} \to 2Fe^{III}}$$

This method of expressing the chemistry shows that electrons are leaving the carbon, which is being oxidized, and joining the iron, which is being reduced.

In the reaction of elemental iron and sulfur to produce iron sulfide discussed above, the sulfur (like oxygen) is considered to be in its reduced -II state. The iron is considered to be in its II oxidation state. An understanding of why oxygen and sulfur in their combined states are sometimes assigned an oxidation state of -II will be gained from Chapter 11 (oxygen and sulfur are both members of Group 16 of the periodic classification, Fig. 2.1). Before oxidation states were used, the term oxidation was restricted to reactions in which elements or compounds reacted with dioxygen, e.g. carbon burning to give carbon dioxide; and the term reduction was restricted to reactions in which elements or compounds reacted with dihydrogen, e.g. a metal oxide, e.g. zinc oxide, ZnO, reacting with dihydrogen to give the metal and water:

$$ZnO + H_2 \longrightarrow Zn + H_2O$$

which may be written in terms of changes in oxidation states of zinc and hydrogen as:

$$\text{or } Zn^{II} + 2H^{0} \longrightarrow Zn^{0} + 2H^{I}$$

2e⁻ (from $2H^0$ to Zn^{II})

The oxidation state concept allows a far greater number of reactions, i.e. those in which oxygen and hydrogen are not participants, to be considered in terms of the transfer of numbers of electrons between constituent atoms of the reactants. For example, when powdered zinc metal (Zn^0) is added to a solution of copper sulfate (Cu^{II}) a reaction occurs:

$$Zn^{0} + Cu^{II} \longrightarrow Cu^{0} + Zn^{II}$$

in which the zinc metal (the reducing agent) is oxidized to its II state, and the copper (the oxidizing agent) is reduced to its zero state. Two electrons are transferred from the zinc(0) to the copper(II) in the process. If a piece of metallic zinc is placed into a blue solution of copper sulfate, the above reaction occurs. Some of the zinc metal dissolves in the solution, copper metal being deposited in its place. If there is sufficient zinc metal to cause the complete reduction of the copper(II), the solution becomes colourless because zinc(II) ions do not absorb visible light.

3

Radiation and radioactivity

Light and the electromagnetic spectrum. Photons. Colour. The effects of radiation on matter. Photosynthesis. Atomic nuclei. Unstable nuclei. Radioactivity. Alpha particle emission. Beta particle emission. Gamma ray emission. Radioactive decay processes. Nuclear fission. Uranium. Plutonium. Nuclear reactors. Nuclear fusion. Health and safety aspects.

3.1 LIGHT, THE ELECTROMAGNETIC SPECTRUM AND PHOTONS

Light is one kind of electromagnetic radiation. The various kinds of electromagnetic radiation are forms of energy that differ only in their wavelength ranges. Everyone who has been warmed by exposure to sunlight knows that radiation is a form of energy. Central heating radiators emit infra-red energy that can be sensed at close range as heat, although they distribute most of their heat by convection as heated air which rises. All the kinds of electromagnetic radiation travel at the very high speed of 299,792,458 metres per second (186,282 miles per second), taking about eight minutes to reach Earth from the Sun.

The two measurements which characterize electromagnetic radiation are frequency and wavelength, usually denoted by the Greek letters ν (nu) and λ (lambda) respectively. The wavelength and frequency of electromagnetic radiation are related by the equation:

$$\lambda \times \nu = c$$

where c is the velocity of light. Fig. 3.1 is a diagrammatic representation of a wave motion and indicates what is meant by wavelength. The wavelength is the distance travelled by the wave in completing one cycle of oscillation. The associated frequency of the wave motion is expressed as the number of cycles per second, the unit for this being the Hertz (Hz), one cycle per second being equal to 1 Hz. The energy of electromagnetic radiation is proportional to its frequency: double the frequency and the energy doubles. This relationship is expressed symbolically by the Planck equation:

$$E = h\nu$$

where E is the energy of the radiation, the factor h is Planck's constant and has the very small value of 6.626×10^{-34} Joule seconds, and ν is the frequency. The symbol $h\nu$ may be used to represent a 'particle' of radiation, known as a **photon.** Photons are described later in this section.

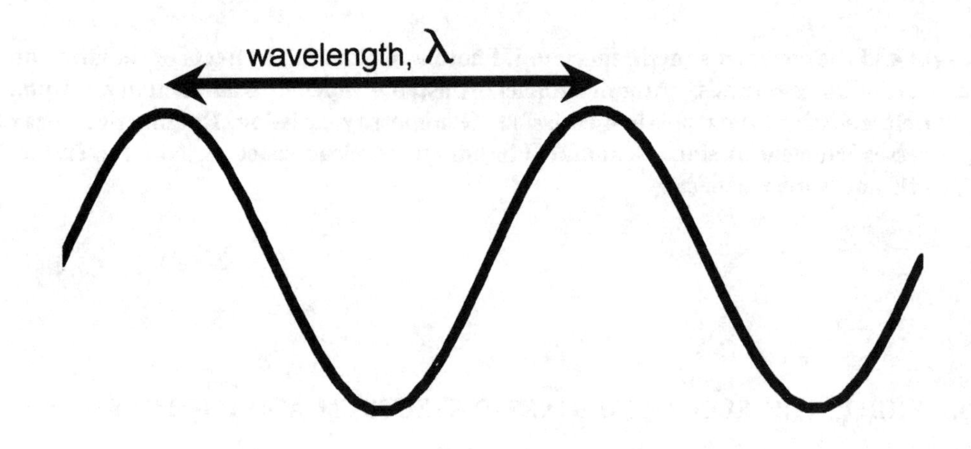

Fig. 3.1 A wave-form showing what is meant by wavelength

The full range of the spectrum of electromagnetic radiation varies from very high energy gamma-rays at one end and very low energy radio waves at the other. The divisions of the spectrum are shown diagrammatically in Fig. 3.2.

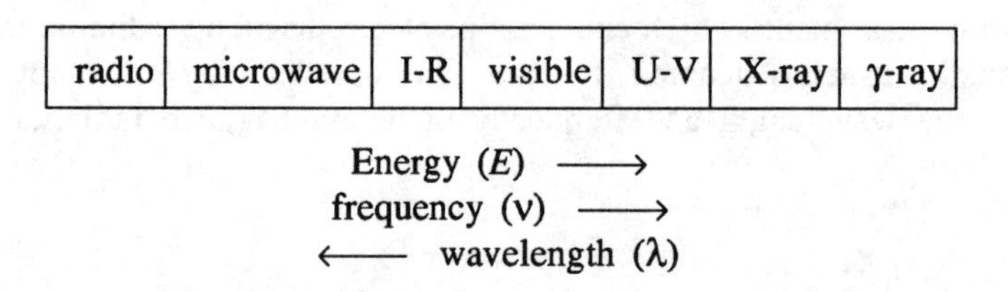

Fig. 3.2 The regions of the electromagnetic spectrum. As indicated by the directions of the arrows, the energy (E) and frequency of radiation (ν) increase from left to right, the wavelength (λ) increasing from right to left. The sizes of the regions are not to scale, having widths sufficient to contain the text

Max Planck (1858-1947) and Albert Einstein (1879-1955) were mainly responsible for our current understanding of radiation and its equivalent energy. For their work on quantum theory (Planck) and the photoelectric effect (Einstein) they were awarded Nobel Prizes for physics in 1918 and 1921 respectively. Planck's work was concerned with the distribution of the intensities of frequencies of continuous radiation emitted by hot solids. The emitted energy is now known as cavity radiation, i.e. that which is emitted by a furnace or a hot oven, and is described in detail in Chapter 4. Planck was able to explain the phenomenon only by using the idea that energy was emitted from vibrating units of the hot solid in discrete 'packets' which he called quanta (singular: quantum). Until then there was a general acceptance of the so-called 'wave theory' of electromagnetic radiation. The basis of that theory was that radiation was a continuous wave motion. The continuous waves on the sea are responsible for the transmission of energy derived from the winds and tides, but do not represent a strict analogy to radiation. Light travels through the vacuum of space where there is nothing to carry waves. The continuous-wave theory was unable to explain the distribution of radiation given out by hot solids. Quantum theory explained the distribution perfectly.

It was still a difficult matter to reconcile the ideas of radiation being a wave motion and also being made up of discrete packets of energy or quanta. Einstein made great progress with the understanding of the problem when he became interested in the photoelectric effect. When light is allowed to shine on a clean metal surface electrons may be released. Such electrons are called photoelectrons because they are produced by light. Einstein's contribution to the elucidation of the photoelectric effect was to consider the quanta of energy to be 'particles' of energy. The interaction of a 'particulate' quantum, sometimes called a photon, with a metal surface became understandable in terms of the energy of the quantum and the minimum energy required to remove an electron from the metal surface. The excess of energy greater than that minimum is converted into the kinetic energy, i.e. the energy of movement, of the photoelectron, causing it to acquire a high speed.

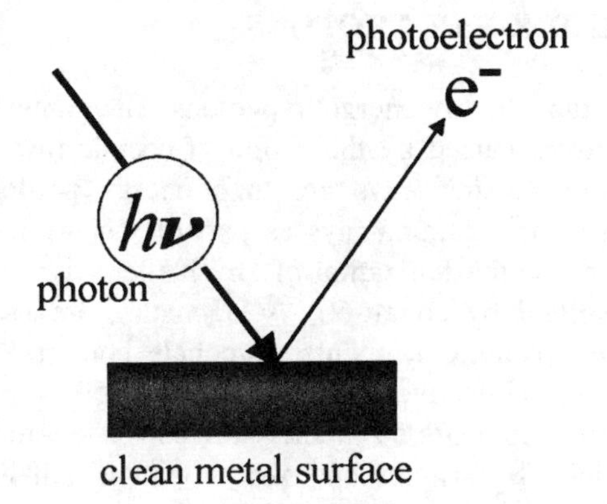

Fig. 3.3 A diagram showing the production of a photoelectron (e^-) when a sufficiently energetic quantum (of energy $h\nu$) of electromagnetic radiation strikes a clean metal surface

Fig. 3.3 shows a diagram of the production of a photoelectron as the result of a photon, indicated by the symbol $h\nu$ which is explained above, possessing sufficiently high energy hitting a metal surface. If the incident photon does not possess sufficient energy to cause the removal of an electron from the metal surface, no emission is observed.

The theory of quantum electrodynamics, as it is now known, was fully modernized by Richard P. Feynman who shared the 1965 Nobel Prize for physics. He showed that there were no fundamental difficulties in considering light, and electromagnetic radiation in general, as being composed of particles of energy.

The modern view of the arrangement of electrons in atoms is that electrons occupy volumes of space within the atom which are called orbitals. The portrayal of an atom as a nucleus with the electrons circulating around in circular or elliptical orbits, very common in publications concerned with the popularization of science, is unhelpful and possibly confusing. The idea is too simple and does not stand the test of scrutiny by modern physical principles. It was first proposed by Niels Bohr (1922 Nobel Prize for physics) and represents an important stage in the development of atomic theory. Bohr realized that, according to the laws of classical physics, an orbiting negatively charged electron should radiate energy continuously as it spiralled into the positively charged nucleus. That this did not happen led him to postulate in 1932 that the electrons could only occupy the fixed, i.e. quantized, levels that were available to them in any given atom. This was a very original and far-reaching piece of thinking and allowed the interactions of radiation with matter to be rationally interpreted. The concept is analogous to the fixed positions that a person may adopt on a flight of stairs, transitions between any two levels being explicable in terms of definite heights rather than the continuous movement of a person on an escalator.

Photons may be classified in terms of their energies and their effects upon matter with which they interact.

3.2 GAMMA (γ) AND X-RAY PHOTONS

Gamma-rays are the most highly energetic photons. The gamma ray range includes high energy rays which are produced by the action of cosmic rays upon the components of the upper atmosphere. Cosmic rays are high energy particles which interact with matter to give high energy gamma rays as part of the mechanism which slows them down. Gamma-rays cause the ionization of any matter with which they interact. Some γ-rays, e.g. those emitted by cobalt-60, $^{60}_{27}Co$, nuclei, are used in cancer therapy to destroy tumour cells. Naturally occurring cobalt consists wholly of the mass-59 isotope which is non-radioactive. If natural cobalt-59 is irradiated with neutrons in an atomic reactor, the mass-59 nuclei are converted to those of mass-60. The cobalt-60 isotope emits β particles and γ-rays with a half-life of 5.27 years. These nuclear changes are explained in Section 3.9.

X-rays were discovered by Röntgen for which he was awarded the 1901 Nobel Prize for physics, the first year in which the prize was presented. X-rays were originally known as Röntgen rays. X-ray photons have lower energies than those of γ-rays, but are still energetic enough to cause ionization of atoms with which they come into contact. The X-rays used by hospitals for diagnostic and therapeutic purposes are produced by allowing a stream of accelerated electrons to hit a copper metal target. The bombarding electrons cause the atoms of copper to be ionized by removing low energy electrons from them. Electrons which are less stable fall down to take the places of the electrons which have been removed and cause the emission of the characteristic X-ray spectrum of copper. Such X-rays are absorbed by the denser parts of the human body, e.g. bones, and by dense substances which are introduced to the body by mouth, e.g. barium sulfate meals, or by injection, e.g. barium sulfate enemas or compounds of iodine by hypodermic injection. Barium (atomic number = 56, RAM = 137) and iodine (atomic number = 53, RAM = 127) are very effective in absorbing X-rays because of their relatively high nuclear charges and masses. The major constituents of the body have low values of atomic number and RAM; carbon (Z = 6, A = 12), oxygen (Z = 8, A = 16), nitrogen (Z = 7, A = 14) and hydrogen (Z = 1, A = 1). These lighter elements do not absorb X-rays effectively.

The high energy photons, gamma-rays and X-rays, are sometimes classified as ionizing radiation since they cause the material with which they interact to lose electrons. They are absorbed indiscriminately by materials and produce a positive ion and an electron: a photoelectron. The latter usually has a sufficiently high energy to cause further ionizations. When the various electrons produced in secondary ionizations have lost most of their energy they react with any suitable molecule to cause chemical reactions. The irradiation of foods to cause sterilization is the subject of some debate. The main component of foodstuffs is water and since ionizing radiation is absorbed indiscriminately the main effects are upon the water content.

The effect of absorbing a photon of gamma radiation upon water is represented by the equation:

$$H_2O \xrightarrow{\gamma} H_2O^+ + e^-$$

the photoelectron causing more ionizations and a track of positive water molecule-ions and electrons. The electrons eventually lose their energy and become aquated or hydrated; they are surrounded by a hydration shell of water molecules:

$$e^- \longrightarrow e^-(aq)$$

The hydrated electrons may react with a water molecule:

$$e^-(aq) + H_2O \longrightarrow H + OH^-$$

to give a hydrogen atom and a hydroxide ion. The positive water molecule-ions decompose to give an aquated proton and a hydroxyl free radical:

$$H_2O^+ \longrightarrow H^+ + OH$$

The hydrogen ions and hydroxide ions neutralize each other by forming water. The irradiated system then contains two very reactive electrically neutral species; hydrogen atoms (H) and hydroxyl free radicals (OH). These may react with other substances, i.e. the rest of the food or bacteria present in the food, or could either react with each other to give water or undergo dimerization reactions to give molecular dihydrogen and hydrogen peroxide (H_2O_2), the latter being a very effective bactericide. A very small dose of gamma radiation is required to sterilize any food.

3.3 ULTRA-VIOLET RADIATION

The ultra-violet region which may directly affect life on Earth is a very narrow band which is that not absorbed by dioxygen and ozone (O_3, trioxygen) in the stratosphere. Even so, it is divided by the manufacturers of sunglasses and sun protection creams and oils into two regions called UVA (wavelengths between 400-320 nanometres) and UVB (320-280 nm). One nanometre (nm) is equal to 1×10^{-9} metres or one one-thousand-millionth of a metre.

Advertisements have appeared which claim that particular sunglasses, sun creams, sun oils and sun lotions absorb the UVC (280-100 nm) region of the ultra-violet spectrum in spite of the fact that none of this radiation reaches the Earth's surface. It is entirely absorbed by the oxygen of the upper atmosphere and the ozone layer in the stratosphere. Claims that some UVC is reaching the Earth's surface are false, although the claims in the advertisements with regard to the absorption of UVC radiation are true. If any UVC did penetrate the ozone layer the sun-protection products would absorb it. The layer would have to be completely destroyed for this to happen and dioxygen would have to lose its capacity for absorbing the UVC radiation. The UVC radiation would only reach the Earth's surface if the atmospheric dioxygen level fell to below 1% from the present level of nearly 21%. If that were to happen there would be no human life left to express any concern about the matter.

The UVA radiation produces a temporary sunburn when allowed to be incident upon human skin. The UVB radiation has a different mode of action and produces tanning which is more lasting, can initiate skin cancer, and should be avoided. Sunburn is avoided either by covering exposed parts of the skin with a lotion containing a substance which absorbs the offending ultra-violet wavelengths or, much more cheaply and sensibly, by staying in the shade and wearing suitable clothing and a large-brimmed hat when shade is not available. It is very important that none of the ultra-violet radiation should enter the eyes.

Unlike the absorption of ionizing radiation, the absorption of ultra-violet radiation is very much dependent upon the absorbing material and is very selective. The ultra-violet radiation falling upon a sun protection oil may all be absorbed by the active component of the oil and none by the oil itself. The oil component acts as a solvent and suitable carrier for the active component.

3.4 VISIBLE LIGHT AND COLOURS

Visible light is electromagnetic radiation with wavelengths between 380 and 720 nm. The wavelength ranges for various colours are given in Table 3.1. They are only approximate because of variations in the human eye which cause different people to have slightly different perceptions of light and colour.

Table 3.1 - The colours of the visible region of the electromagnetic spectrum and their wavelength ranges in nanometres(nm)

colour	wavelength range/nm
violet	400-420
blue	420-490
green	490-530
yellow	530-590
orange	590-640
red	640-720

The visible region of the spectrum affects the eye in its different ways and gives humans the sense of colour which is generally appreciated. It enhances communications and allows painted art to exist and be appreciated. The white light that reaches Earth from the Sun, whether it be the Sun's direct rays or the diffuse light that allows sight to operate in cloudy conditions, interacts with the materials and substances that it is incident upon in three ways. The division of visible light into the spectrum of rainbow colours, red, orange, yellow, green, blue and violet is well known. In addition to this range of colours are some mixtures which have separate names, e.g. purple (red and violet), indigo (violet and blue) and turquoise (blue and green).

The light incident upon an object may be reflected, transmitted or absorbed. If light is largely reflected from an object it will have a white or metallic appearance. If light is largely transmitted by an object, e.g. a glass of water, both glass and contents being transparent, the object will be colourless. If light is completely absorbed by an object the object will appear to be black. Those objects and materials in which there is some dependence upon wavelength of the three properties of reflexion, transmission and absorption will appear to have colours of various shades and intensities. A highly polished smooth chromium metal surface is known to be a good reflector and does not differentiate between the colours. If an object absorbs visible light selectively its perceived colour is complementary to that of the fraction which is absorbed. The 'artist's wheel' shown in Fig. 3.4 is a summary of colours and their complements. A copper metal surface is also a good reflector but there is an absorption of the blue end of the spectrum which gives copper its characteristic reddish sheen. A red brick has a rough surface and reflects much less efficiently than a polished copper surface, but does absorb the blue end of the spectrum so that it appears to be a dull red colour. Red wine also absorbs the blue end of the spectrum but transmits some of the remainder of the visible light as well a reflecting another fraction. Leaves and grasses generally absorb the red and blue parts of the spectrum, which are essential to the process of photosynthesis, and reflect the remainder to cause their green and greenish-yellow colours. Life without colours would be dull indeed.

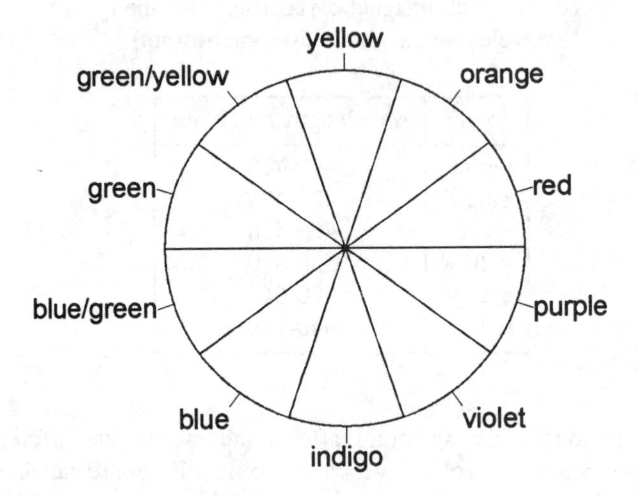

Fig. 3.4 A diagram showing pairs of complementary colours; a material exhibits a colour which is diametrically opposite to that which it absorbs

3.5 PHOTOCHEMISTRY AND PHOTOSYNTHESIS

Absorption of visible and ultra-violet radiation causes the electronic energy of the absorbing molecule to increase in a quantized step-wise manner. The production of an excited molecular state may be written as:

$$M + h\nu \longrightarrow M^*$$

There are also changes in the vibrational and rotational energies of the molecule. The electronically excited states initially produced may have several possible fates. The next collision that the excited molecule undergoes may result in the sharing of the excitation energy between the two colliding molecules:

$$M^* + X \longrightarrow M + X$$

where X represents any molecule (including M) with which the excited molecule (M^*) collides. The system would become a little warmer in consequence. Such a consequence is known as collisional deactivation and is the most common fate of an electronically excited state. A much less common fate of an excited state is to lose its excitation energy by emitting a photon of light, a process known as fluorescence:

$$M^* \longrightarrow M + h\nu'$$

Usually the wavelength of fluorescent radiation is longer than the wavelength of absorbed radiation, i.e. ν' is lower than the exciting frequency, ν. It is common to observe visible blue fluorescence from a molecule which absorbs specifically in the ultra-violet region. This is because the excited state of a molecule loses its excess of vibrational energy before it loses the fluorescent photon which, in consequence, has a lower energy than the photon which initiated the electronic excitation. Fluorescent molecules are usually rather rigid molecules which store their energy of excitation for a long enough time to allow fluorescence to compete with collisional deactivation. The 'blue whiteners' contained by washing powders are compounds which absorb ultra-violet radiation and emit blue fluorescence, thereby giving a fresher appearance to the clean white clothes. The fluorescence is intensified and becomes more obvious when the white materials are subjected to ultra-violet radiation from the lights such as those used in discos.

A third possible fate of an electronically excited state is decomposition either to give two stable molecules or to give two free radicals. Irradiation of the hydrogen peroxide molecule with ultra-violet radiation causes it do dissociate into two hydroxyl free radicals:

$$H_2O_2 \xrightarrow{UV} 2OH$$

The hydroxyl radicals attack other hydrogen peroxide molecules by removing a hydrogen atom to produce water and another free radical, which is a fragment of the hydrogen peroxide molecule, called hydroperoxyl, HO_2:

$$OH + H_2O_2 \longrightarrow H_2O + HO_2$$

The hydroperoxyl radicals may destroy other hydrogen peroxide molecules according to the equation:

$$HO_2 + H_2O_2 \longrightarrow H_2O + O_2 + OH$$

the hydroxyl radical continuing the chain decomposition of the hydrogen peroxide. Such chains are terminated if the two chain carriers meet and react according to the equation:

$$HO_2 + OH \longrightarrow H_2O + O_2$$

The overall effect of the ultra-violet radiation is to convert hydrogen peroxide into water and dioxygen. The presence of free radicals in the decomposition of hydrogen peroxide is essential to the bleaching and sterilization actions of its aqueous solutions. In the laboratory, photochemistry is commonly carried out by using a mercury vapour lamp. The lamps are similar to those which are used in street lights, the glass covering of street lights preventing emission of ultra-violet radiation.

The study of the chemical changes initiated by the absorption of photons in the visible and ultra-violet regions form the basis of photochemistry. The most important photochemistry is photosynthesis upon which all life depends. The overall process of photosynthesis may be written as the simplified equation:

$$xCO_2 + xH_2O \xrightarrow{\text{light}} xO_2 + (CHOH)_x$$

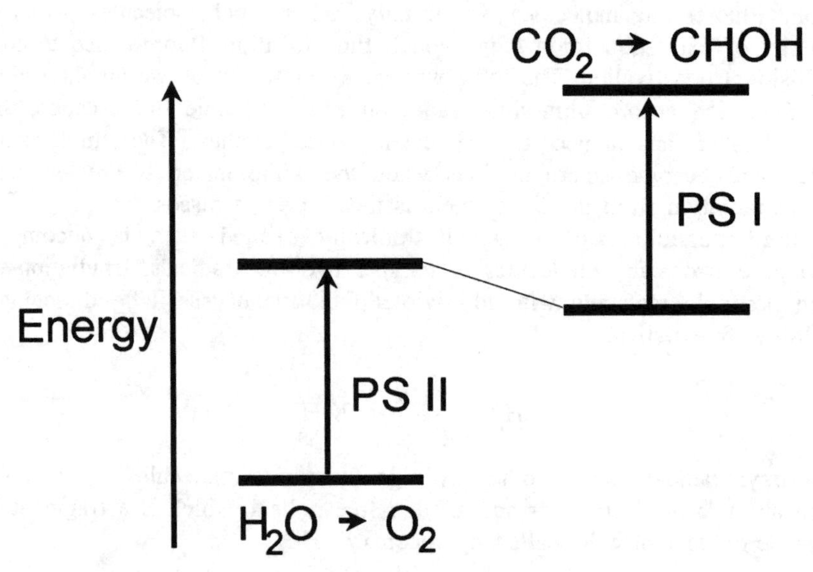

Fig. 3.5 A diagram showing the energy changes and reactions associated with photosynthesis; the formula CHOH (i.e. C + H_2O) is representative of the composition of the carbohydrate products; PS I and PS II are the two photo-systems by which the Sun's light is used to convert carbon dioxide into carbohydrate and to oxidize water to dioxygen

The organic products (with various values of x, a value of x = 6 corresponding to the formation of glucose) are carbohydrates from which all the plant's structures are formed in subsequent reactions. The process requires the presence of two types of chlorophyll, called *a* and *b*, which are complex molecules containing magnesium. Chlorophyll and other complex molecules are contained by green leaves and are responsible for the absorption of visible radiation with wavelengths in the red and blue regions. The leaves have a capacity for using the absorbed energy to make carbohydrate synthesis possible. The processes participating in photosynthesis are numerous and complex. Essentially there are two photo-systems, one in which the absorbed radiation is used to convert carbon dioxide into carbohydrates, called photosystem I (PS I), the other in which absorbed radiation is used to convert water into oxygen, called photosystem II (PS II). The associated energy changes are shown diagrammatically in Fig. 3.5. It is known that all the oxygen atoms in the synthesized carbohydrates originate in the assimilated carbon dioxide and that all the oxygen atoms of the evolved oxygen originate in water molecules. The two photo-systems are connected by 'dark' or thermal reactions, which are normal chemical reactions, and consist themselves of series of such reactions in addition to the initiating photochemical stages. This may be understood generally in terms of the two equations:

$$2H_2O \xrightarrow{\text{PS II}} 4H^+ + O_2 + 4e^-$$

(water oxidation reaction)

$$CO_2 + 4H^+ + 4e^- \xrightarrow{\text{PS I}} CH_2O + H_2O$$

(carbon dioxide reduction reaction)

The hydrogen ions and electrons, which participate in further reduction/oxidation reactions, shown in the two half-reactions above, pass between the two photo-systems in a series of complex processes which include protein molecules with iron, sulfur and manganese atoms having important roles. The sum of the two half-reactions represents the overall process of photosynthesis. In effect the electrons produced in PS II lose some of their energy in travelling to the site of PS I where they receive another photochemical boost, i.e. an absorption of photons, to enable the reduction of carbon dioxide to occur. The overall process of photosynthesis is very inefficient in that only 1% of the radiation incident upon green vegetation is actually used in the reactions. In theory, three photons of wavelengths in the 680-700 nm band have a combined energy which is sufficient to promote the reaction, but in practice eight photons are required. In molar terms this means that the assimilation of one mole of carbon dioxide requires eight moles of photons of appropriate energy.

The reverse reactions, those in which carbohydrates are oxidized (sometimes burned), are very exothermic and represent the normal direction for a spontaneous process. That photosynthesis is the reversal of the burning of glucose, or of a polysaccharide such as starch, emphasizes the importance of the Sun's energy as the basis of life.

The possible use of solar energy directly instead of burning non-renewable resources, i.e. natural gas, petroleum and coal, is discussed in Chapter 6.

3.6 INFRA-RED RADIATION

Infra-red radiation is that which emanates from gas, wood or electric fires and from hot bodies in general, causing the sensation of heat upon the human skin. The energy of the radiation causes molecules in the skin to vibrate more energetically which gives the sensation associated with a rise of temperature.

When infra-red radiation interacts with molecules their vibrational energy is increased. Vibrational energy is quantized so that only photons with characteristic frequencies of infra-red radiation are absorbed by a particular molecule. A diatomic molecule can vibrate in only one way, the bond length becoming longer and shorter than its average value. In order to interact with infra-red radiation the vibrational motion must be associated with a change in the polarity of the molecule. Diatomic molecules like dioxygen and dinitrogen are non-polar and cannot interact with infra-red radiation. Diatomic molecules with different atoms such as carbon monoxide, CO, are polar and can increase their vibrational energy by absorbing photons of infra-red radiation. Carbon monoxide absorbs radiation at a wavelength of 4.6 micrometres which causes the molecule to vibrate more energetically. Hydrogen chloride absorbs at a wavelength of 3.5 micrometres in a similar transition. The difference in the wavelengths absorbed by the two molecules is due to the differences in bond strength and in the masses of the vibrating atoms. Molecules with more than two atoms, known as polyatomic molecules, can undergo both stretching and bending vibrations and providing that the individual vibrations cause a difference in the molecule's polarity they may be excited by absorbing the appropriate wavelength of infra-red radiation.

3.7 MICROWAVE RADIATION

Microwave radiation is used in the home for thawing frozen food and for cooking. The ovens are usually tuned to produce radiation with a wavelength of 12.24 centimetres, the corresponding frequency being 2.45 gigahertz. Microwave radiation interacts with ions, e.g. Na^+ and Cl^-, polar molecules, e.g. water, and molecules containing polar groups, e.g. sugars with OH groups, to increase the energy of the absorbing material. Ions and polar molecules are accelerated, the polar molecules also having their rotational energies increased. The increased rotational energy is eventually converted to translational energy of the components of the irradiated system. This causes the temperature of the system to increase. Microwave irradiation of molecular systems does not initially cause chemical change, which however may be caused by the increase in temperature that follows the conversion of the microwave energy into the vibrational and kinetic energy of the absorbing system. Food is cooked more quickly in a microwave oven than in a conventional one because microwave radiation is more penetrative than that in the infra-red region. In a conventional oven the heat reaches the interior of the food by conduction from the surface, the centre of the food being the last to be cooked. Microwave cookery ensures a much more even distribution of the energy throughout the food from its surface to its interior, an even radial distribution being achieved by rotating the food. Because the microwave radiation is absorbed by bones it is essential to turn over bone-containing foods at the half-way stage to ensure complete cooking.

The radar beams used in radar-guns, and in detectors generally, make use of wavelengths in the microwave region.

3.8 RADIO WAVES

Beyond the microwave region is the very extensive portion of the electromagnetic spectrum which is used for radio and television communications.

Photons in the radio wave region of the spectrum possess insufficient energy to cause changes in molecular systems. Radio frequency radiation can interact with metallic objects such as the copper wire from which radio aerials are constructed. It can also cause minute changes within atomic nuclei. There are some atomic nuclei, e.g. ^{1}H and ^{13}C, which absorb radio frequencies when subjected to a high magnetic field. The magnetic field allows certain nuclei, i.e. those with magnetic properties, to adopt one or other of a small number of quantized positions with regard to the applied field. It is as though each of these nuclei were acting as little bar magnets with north and south poles. A proton has two allowed, i.e. quantized, positions of its nuclear magnet; it is either parallel to or opposed (anti-parallel) to the north and south poles of the applied field. The nuclei which are exposed to radio frequencies in the presence of a magnetic field undergo resonance between the two allowed positions and absorb at precise frequencies in their nuclear magnetic resonance (nmr) spectrum. Nmr spectroscopy is widely used in chemical research for the identification and structure determination of molecules and in the diagnostic whole-body scanning machines used in hospitals.

3.9 RADIOACTIVITY

The study of radioactivity was important in the development of the theory of atomic structure. Radioactivity is a property exhibited by unstable nuclei. It was discovered in 1896 by Henri Becquerel who noticed that uranium salts could affect a photographic plate through a layer of black paper. Other radioactive substances were found and eventually, the element radium was isolated from pitchblende by Marie Curie. She was awarded the Nobel Prize for chemistry in 1911 for the discovery of radium and polonium. For their work on radioactivity, Marie Curie, her husband Pierre, and Becquerel, shared the Nobel Prize for physics in 1903. It was soon realized that radioactivity in general consists of three kinds of rays emanating from various radioactive materials. The rays were classified as alpha (α), beta (γ) or gamma (γ), depending upon whether their electrical charges were positive, negative or neutral respectively, and their capacities to pass through materials of various thicknesses.

Alpha-rays were identified as helium nuclei, $^{4}_{2}He^{2+}$, and are now properly described as alpha-particles. Beta-rays were identified as electrons and are now properly called beta-particles. Gamma-rays are a form of electromagnetic radiation, as described in Section 3.2.

The relative difficulties of stopping, i.e. by absorption, alpha- and beta-particles, and gamma-rays, are shown in Fig. 3.6. Alpha-particles are easily absorbed by paper or thin metal foil such as the aluminium foil used in cooking and food protection. They have a very limited range in air, typically between three to eight centimetres. Beta-particles are more penetrative than α-particles. They are totally absorbed by a one millimetre thickness of aluminium. Gamma-rays are extremely penetrative, requiring a thickness of around thirty centimetres of lead or one metre of concrete to prevent their passage.

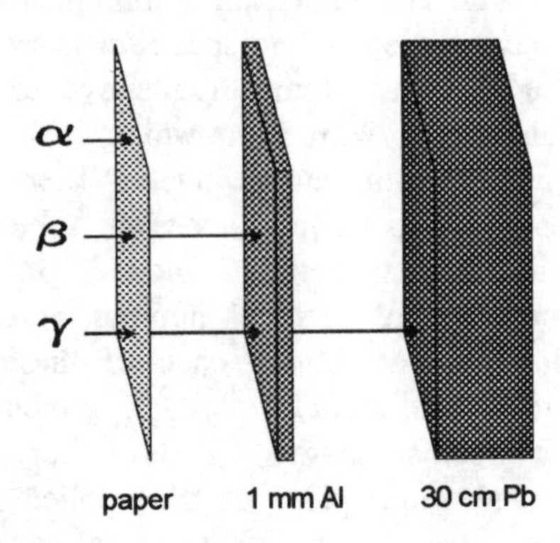

Fig. 3.6 A diagram showing the relative difficulties of stopping α and β particles and γ rays; α particles will not pass through a piece of paper, β particles are absorbed by a one millimetre thick sheet of aluminium and the more penetrative γ-ray photons are only stopped by a 30 centimetre thick block of lead

Gamma-ray emission is now known to be a secondary radioactive event, occurring after either α or β emission from individual atoms has taken place. A radioactive nucleus decays either by α or β emission, resulting in the formation of a new nucleus which may be referred to as a daughter nucleus. In the field of radioactivity all nuclear parents produce daughters. Some daughter nuclei are produced in an excited energetic state. When this new excited nucleus loses its excitation energy to give the final product nucleus the excitation energy is emitted as a γ-ray photon. The processes may be illustrated in general terms by the equations:

Primary process:

$$\text{Parent nucleus} \longrightarrow \text{Excited daughter nucleus} + \alpha \text{ (or } \beta)$$

Secondary process:

$$\text{Excited daughter nucleus} \longrightarrow \text{Daughter nucleus} + \gamma$$

The naturally occurring radioactive isotopes of the elements are almost entirely confined to those with atomic numbers greater than 83, no isotope with an atomic number less than 84 and a mass number, i.e. the nearest whole number to the exact mass, less than 209 being an α-emitter. Some naturally existing isotopes of lighter elements are radioactive, almost entirely as β-emitters: these are ^{14}C, ^{40}K,

^{50}V, ^{87}Rb, ^{115}In, ^{123}Te, ^{130}Te, ^{138}La, ^{144}Nd, ^{147}Sm, ^{148}Sm, ^{149}Sm, ^{176}Lu, ^{174}Hf, ^{180}Ta, ^{187}Re, ^{186}Os and ^{190}Pt. Of these exceptions only the ^{87}Rb (27.8%), ^{115}In (95.7%), ^{130}Te (33.9%), ^{144}Nd (23.8%), ^{147}Sm (15%), and ^{187}Re (62.6%) nuclei contribute considerably to the naturally occurring element, their respective abundances being indicated in parenthesis. They are all beta-particle emitters with half-lives (defined below) of at least 10^{11} years.

All radioactive decay processes are governed by the same piece of mathematics. This is that the rate of decay of any isotope is proportional to the number of parent nuclei of that isotope present in the sample. Independently of their number, the decay of one half of the parent nuclei of any one isotope occurs in a time known as the half-life. The half-life of a given isotope is an important characterizing property which allows the identity and amount of the isotope present in any sample of material to be measured. The decay of a mixture of isotopes, with their differing half-lives, may be resolved into contributions from the individual components, allowing them to be identified. The half-life of a particular isotope is absolutely characteristic of that isotope, being independent of temperature and the physical and chemical states of the element. The half-life of the $^{238}_{92}U$ nucleus is 4.46 x 10^9 years (4.46 billion years), a figure which applies to uranium-238 in its elemental metallic state or in compounds which might be solids, liquids, solutions or gases.

One important application of the measurement of radioactive decay rates is that known as radiocarbon dating. Bombardment of atmospheric nitrogen with neutrons (symbol: $^{1}_{0}n$), a constituent of cosmic rays which originate in outer space, produces ^{14}C nuclei:

$$^{14}_{7}N + ^{1}_{0}n \longrightarrow ^{14}_{6}C + ^{1}_{1}H$$

Carbon-14 is a β emitter and decays to give nitrogen-14:

$$^{14}_{6}C \longrightarrow ^{14}_{7}N + e^{-}$$

The half-life of the process is 5668 years. The combined processes of production and decay of carbon-14 cause atmospheric carbon dioxide to have a very small, but constant, carbon-14 content. By the processes of photosynthesis the carbon from atmospheric CO_2 is incorporated in all living matter. While an organism, animal or vegetable, is still alive its carbon-14 content, relative to that of its stable carbon-12, is the same as that of the CO_2 in the atmosphere and is continually replenished by the processes of life, e.g. breathing and eating, in the case of animals, photosynthesis in the case of vegetation.

When an organism dies there is no longer any mechanism for it to exchange its carbon content with the carbon in its environment. The carbon-14 content of the dead organism decreases as time progresses. The carbon-14 content decreases by 1% every 82 years. By careful measurement of the decay rate of the carbon-14 of a dead object, its age may be estimated. Usefully accurate estimations of ages between 1000 and 20,000 years for dead organic matter can be made. Recently, the *Shroud of Turin*, supposedly wrapped around the dead body of Jesus Christ, was carbon-14 dated and shown to be only around 500 years old. This is a good example of one significant scientific observation being sufficient to destroy a theory. The origin of the image on the shroud is still in doubt. In Italy, at that time, there lived many artists, one by the name of Leonardo da Vinci.

The main block of radioactive isotopes which consist of the heavier elements, i.e. those with Z greater than 83 and A greater than 209, exhibit all types of radioactive decay but they all possess characteristically different half-lives. Uranium and plutonium are by far the more important elements in this region, the former existing naturally in the Earth's crust while the latter's natural abundance is either extremely small or possibly zero.

3.10 URANIUM

The abundance of uranium in the Earth's crust is 2.3 grams per tonne. The element became commercially important in 1939 when Meitner, Hahn and Strassman discovered that the nucleus could undergo decay by fission. The world production of uranium in 1990 amounted to 47,219 tonnes of contained metal, the distribution of the production being shown in Fig. 3.7. The value of the production was $2.2 billion.

The consumption of uranium, based upon reactor requirements in 1990, was 50,200 tonnes of contained metal. Of that total, 86% was used by the Western countries with the distribution given in Table 3.2.

Table 3.2 - The distribution of uranium consumption by Western countries in 1990

group/country	% consumption
USA	34.5
E.E.C.	34.5
Japan	13.5
Others	16.5

It has been estimated that there are sufficient reasonably assured resources of uranium to last for fifty eight years at the present rate of consumption. The distribution of uranium resources is shown in Fig. 3.8.

Uranium occurs as the mineral known as uraninite or pitchblende, UO_2, and an oxide with the formula, U_3O_8. There are some minerals, such as carnotite $K_2(UO_2)_2(VO_4)_2.3H_2O$, which contain the uranyl ion, UO_2^{2+}, which is an important feature of uranium chemistry. The concentrated ores are roasted, to oxidize all the uranium to its VI state, and then treated with a mixture of manganese dioxide and sulfuric acid which ensures that the uranium is converted entirely to its VI state which is soluble in sulfuric acid solution. Dioxygen and manganese dioxide are the oxidants. The uranium content of the solution is removed by precipitating it as ammonium diuranate, $(NH_4)_2U_2O_7$, by neutralization. The precipitate is known as yellow cake because of its strikingly characteristic colour which is practically identical to that of elemental sulfur. The effect of heating the yellow cake is to convert it to the oxide U_3O_8 which is then reduced to UO_2 with hydrogen, converted to the tetrafluoride, UF_4, by reaction with hydrogen fluoride, the metal being produced by the reduction of the tetrafluoride with elemental magnesium:

$$UF_4 + 2Mg \longrightarrow 2MgF_2 + U$$

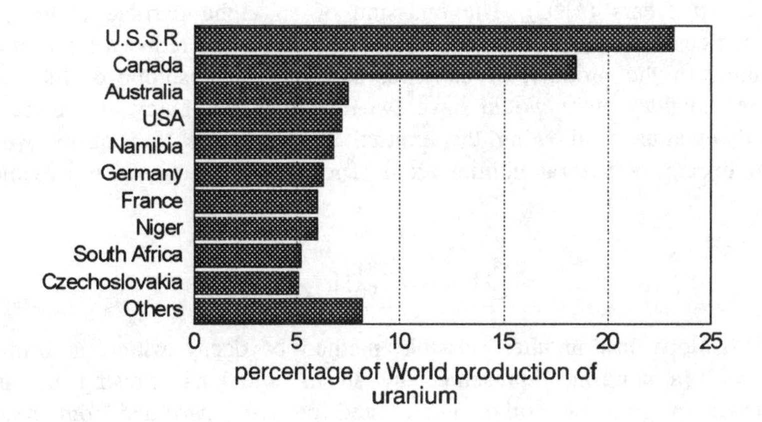

Fig 3.7 The distribution of World production of uranium in 1990

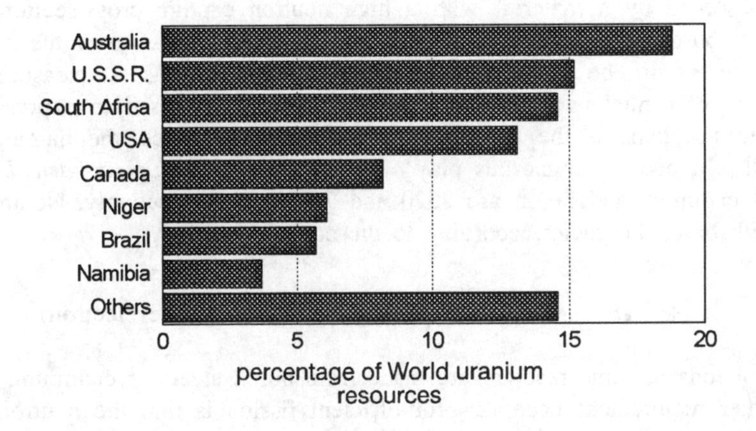

Fig. 3.8 The distribution of World uranium resources

The main use of uranium is in nuclear reactors which produce heat to form steam for the running of electricity generators.

3.11 NUCLEAR FISSION REACTORS

Naturally occurring uranium consists of two isotopes with mass numbers 235 and 238, the former being present to the extent of only 0.72%. Both isotopes are radioactive in that they are alpha particle emitters having half-lives of 7 x 10^8 years (^{235}U) and 4.5 x 10^9 years (^{238}U). The emission of an alpha particle ($^4_2He^{2+}$) from the uranium nucleus reduces its mass number by four units and reduces the atomic number by two units so the product, or daughter nucleus, is an isotope of thorium in both cases. The daughter atom would have two electrons too many and these would be transferred, eventually allowing the emitted alpha particle to acquire two electrons so that it became a neutral helium atom. The overall change is represented by the equation:

$$^{238}_{92}U \longrightarrow ^{234}_{90}Th + ^{4}_{2}He$$

The ^{235}U isotope has another possible method of decay which is initiated by a collision with a neutron, represented by ^1_0n in equations, causing the nucleus to split asunder to give two other nuclei and between two and four neutrons, for example:

$$^{235}_{92}U + ^{1}_{0}n \longrightarrow ^{138}_{53}I + ^{95}_{39}Y + 3^{1}_{0}n$$

A range of nuclear products result from the various fission processes which occur and the over-production of neutrons causes a branching chain reaction, i.e. as happens in a nuclear explosion, to continue unless it is prevented by the absorption of a sufficient fraction of the neutrons by suitable means. The excess of neutrons may be absorbed by a material with a high neutron capture cross-section such as cadmium or boron. The neutron capture cross-section is a measure of the size that a nucleus appears to be to an approaching neutron and is a measure of the effectiveness of a nucleus in capturing a neutron. The unit used to express neutron capture cross-sections is the barn or 1 x 10^{-28} square metres, the naming of this very small unit being a humorous play on the phrase '*as big as a barn door*'! The values for cadmium and boron are 2520 and 760 barns respectively. Neutrons react readily with boron-11 nuclei according to the equation:

$$^{1}_{0}n + ^{11}_{5}B \longrightarrow ^{12}_{6}C + ^{0}_{-1}e \text{ (}\beta\text{ particle or negative electron)}$$

The control rods in some reactors are made from boron steels or cadmium.

Another requirement necessary for efficient fission is that the neutrons should be thermalized, i.e. slowed down until they have thermal energies appropriate to the reactor temperature rather than the high energies which they possess upon emission from the fission process. Fast neutrons freshly emitted in the fission process have too much energy to be efficiently captured by uranium-235 nuclei (neutron capture cross-section = 95 barns) with which they collide. Materials which slow down neutrons are called moderators and possess very small neutron capture cross-sections. Typical moderators are graphite and deuterium oxide, carbon having a neutron capture cross-section of 3.5 millibarns, those of deuterium and oxygen being 0.52 and 0.28 millibarns respectively.

The uranium-238 nucleus is not fissionable, so it is advantageous to use uranium enriched with the 235-isotope in most forms of nuclear reactors. The 238 nucleus has a neutron capture cross-section of only 2.7 barns, but this has important consequences in the manufacture of plutonium. The form in which the uranium is used varies with the reactor design. Early reactors used the metal but it is now more common to use either the dioxide, UO_2, or the carbide, UC_2, the latter being used mainly for the highly compact reactors which provide power for nuclear submarines.

There are various reactor designs. The thermal reactors employ a moderator which is either graphite or water, the generated heat being extracted with a coolant which is either carbon dioxide, helium or water. Thermal reactors are so-called because they depend on the use of slowed down neutrons which have ordinary thermal energies. The fuel may be either enriched with uranium-235 or used with the natural distribution of the two isotopes. The fast reactors do not use moderation (they use fast neutrons) and use the plutonium generated in the thermal reactors as fuel. The coolants used can either be liquid metallic sodium or helium gas.

When uranium-235 is used as the fuel in a reactor the amount of energy generated by the fission process is equivalent to that produced by burning 186 tonnes of coal for every gram of the isotope used. If unenriched fuel is used the coal equivalent of one gram is 1.3 tonnes. Those figures would indicate that nuclear energy might be expected to be very cheap, reminiscent of the hyperbole contained in the quotation '*it [electricity] will be so cheap that we won't be able to meter it*'. We now know that electricity generated from nuclear power stations is not as cheap as that from conventional stations. This, in part, is because of unexpected engineering difficulties, the exceedingly high safety standards imposed, and the high cost of decommissioning a station when beyond its sell-by date. That power can be generated from nuclear reaction at all would have surprised Lord Rutherford who said '*anyone who says power can be generated by splitting the atom is talking moonshine*'. Both the over-optimistic and the over-pessimistic statements were uttered by scientists which shows that such people can be far off the mark in predicting the future.

Some of the neutrons concerned in the fission of uranium-235 in a moderated reactor are captured by the uranium-238 nuclei present in the isotopic mixture. A nuclear reaction occurs which may be written as:

$$^{238}_{92}U + ^{1}_{0}n \longrightarrow ^{239}_{92}U + \gamma$$

in which a new isotope of uranium is produced together with a gamma-ray photon which allows for the change in energy of the particles in the new nucleus. The uranium-239 nucleus is quite unstable and emits a β-particle in decaying to yield an isotope of neptunium:

$$^{239}_{92}U \longrightarrow ^{239}_{93}Np + \beta^-$$

The neptunium nucleus is very unstable and emits a β-particle to give the mass 239 isotope of plutonium:

$$^{239}_{93}Np \longrightarrow ^{239}_{94}Pu + \beta^-$$

The nuclear fuel becomes loaded with plutonium as well as with the fission products of uranium-235 and when appropriate is reprocessed. The plutonium-239 isotope is fissionable and if a fast reactor is set up with the fuel being surrounded by uranium-238 the neutrons from the fission of the plutonium may be used to generate even more of the fuel. This is the basis of the breeder reactor in which more fuel is prepared than is used. This is because three neutrons are produced per fission reaction, all of which can eventually lead to the formation of a new plutonium-239 nucleus. This offers the prolongation of the usage of natural uranium in energy generation but, as yet, there has been no commercialization of the breeder reactor.

There are about 415 nuclear fission reactors currently operating in the world, excepting the small reactors which are used in ships and submarines. Fifty eight percent of these are pressurized water reactors which use pellets of UO_2 with a 3% uranium-235 enrichment as fuel and are moderated and cooled by ordinary liquid water. They are pressurized to prevent the water coolant from boiling at the elevated running temperature. They are used mainly by five countries; USA (74 reactors), France (52), U.S.S.R. (24), Japan (19) and Germany (14). Twenty one percent of the reactors are of the non-pressurized boiling water type which use UO_2 pellets with a 2.2% enrichment of uranium-235 with water acting as both moderator and coolant. The countries which mainly use this type of reactor are USA (37 reactors), Japan (21), Sweden (9) and Germany (7). The third type of reactor used is the carbon dioxide gas cooled, graphite moderated, variety which are either the older Magnox reactors or the more recent advanced (AGR) gas cooled version. The Magnox reactors use unenriched metallic uranium rods with a cladding of a magnesium/aluminium alloy. The AGR's use UO_2 pellets with a 2% enrichment of uranium-235 as fuel. This design (9% of the total reactors used) is used almost entirely (36 out of a total of 39 reactors) by the U.K. Seven percent of the reactors are of the pressurized heavy water (PHW) type which use UO_2 pellets of natural isotopic composition with heavy water (deuterium oxide, D_2O) as both moderator and coolant. The main user of this type of reactor is Canada with 20 out of the 29 reactors in use, India operating five. The fifth type of reactor is one which uses a graphite moderator and is cooled with ordinary water. There are twenty of these reactors (LWG, light water/graphite) in operation, all of them in the former U.S.S.R. They are the same as the reactor which melted down in Chernobyl. There are only four fast breeder reactors operating for research and development purposes, none being used for commercial electricity generation. They are situated in France (two), the U.K. (closed down in 1994) and the C.I.S.

Seven countries operate 77% of the reactors in the world. The numbers of reactors that each of these countries operate are shown in Fig. 3.9 together with the percentage of electricity generated by their nuclear reactors.

There is considerable evidence that natural fission of ^{235}U took place in the rich deposit of uranium ore at Oklo in the Gabon Republic in West Africa. Around 1.8 billion years ago the fraction of ^{235}U in natural uranium has been estimated to have been about 3% as compared to the present day value of 0.72% for the majority of deposits. The half-life for the decay of ^{235}U is about 6.34 times shorter than that for ^{238}U, so as time passes the natural material becomes depleted in the lighter isotope. The three percent level of U-235 is that which is produced for most commercial fission reactors, as described above. In the particularly high concentration of the uranium ore in the Oklo deposit, it was possible for natural fission of the lighter isotope to occur.

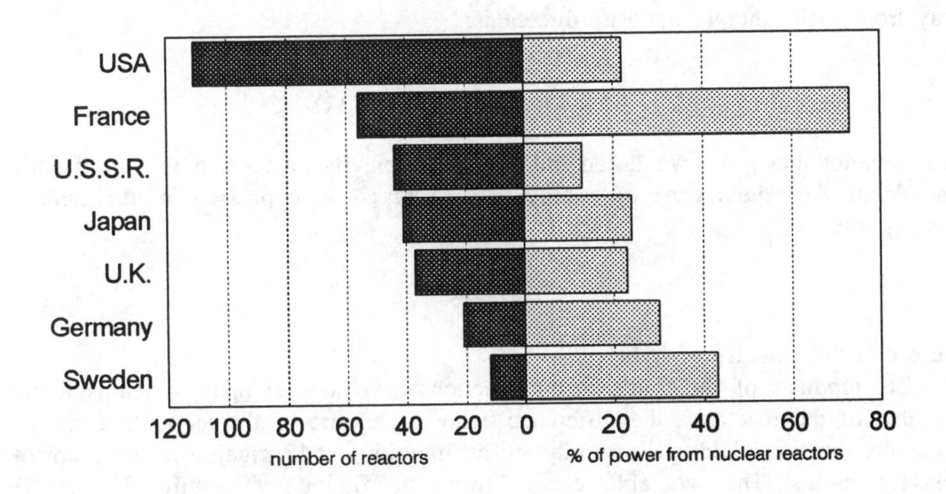

Fig. 3.9 The distribution of nuclear reactors amongst the major nuclear power generating countries and the percentage of their electrical power generated by nuclear reactors

It is possible that the fission process continued for many thousands of years, producing about six tonnes of fission products and causing the depletion of the ^{235}U to its present day level of 0.44% which is significantly lower than any other deposit in the world. The normal level of U-235 in uranium mineral deposits is 0.77%. The distribution of fission products in the Oklo deposit is that which is expected, and is very compelling evidence that natural fission did occur. It is quite possible that other natural fission reactors have been in operation.

3.12 NUCLEAR FUSION PROCESSES

There are difficulties associated with fission reactors connected with the disposal of the highly reactive waste products and which add to the overall cost of the process. Another possibility of harnessing a nuclear reaction as an energy source is nuclear fusion. This is the basis of the Sun's energy, the overall reaction being the fusion of four protons in a multistage process to produce one helium nucleus and two β^+ particles:

$$4\,{}^{1}_{1}\mathrm{H} \longrightarrow {}^{4}_{2}\mathrm{He} + 2\beta^{+}$$

where the beta-plus particles are positive electrons (positrons) that have identical properties, but opposite charge, to the usual negative electron which is so chemically important. The production of a β^+ particle does happen in some radioactive decay products on Earth, especially with synthetic nuclei, the particle reacting with the first electron it meets. The result of the meeting of oppositely charged electrons is their complete conversion into two gamma-ray photons travelling away from each other in opposite directions:

$$\beta^{+} + \beta^{-} \longrightarrow h\nu(\gamma) + h\nu(\gamma)$$

such radiation being known as annihilation radiation. Its production is a very good example of the equivalence of energy, E, and mass, m, expressed by the famous Einstein equation:

$$E = mc^2$$

where c is the velocity of light.

The products of the Sun's fusion reactions have a total mass which is lower than that of the reactants, the difference being a measure of the amount of energy which the reaction yields. The results are an impressive 645 gigajoules per gram of hydrogen atoms. The available energy from the fission of uranium-235 is 81 gigajoules per gram of ^{235}U atoms. The fusion reaction which is likely to be the one used in controlled reactors on Earth is that between deuterium and tritium nuclei:

$${}^{2}_{1}\mathrm{H(D)} + {}^{3}_{1}\mathrm{H(T)} \longrightarrow {}^{4}_{2}\mathrm{He} + {}^{1}_{0}\mathrm{n}$$

the output of which is 844 gigajoules per gram of deuterium or 562 gigajoules per gram of tritium. If the fusion reaction could be controlled on Earth it would serve as a source of energy with an almost inexhaustible fuel reserve. One of the two main difficulties associated with the project is the attainment of sufficiently high temperatures (ten million K) to give the reactant nuclei enough kinetic energy to overcome the electrostatic repulsion associated with two positive particles colliding together. The other difficulty is the impracticality of finding a vessel in which the reaction could occur without the vessel being vaporized by the plasma, the term used to describe the state of nuclei and electrons which is produced at such high temperatures. The high temperatures can be achieved for very short times but it seems to be the containment problem which is yielding very slowly to the research and engineering efforts being expended. The current research efforts are concerned with the containment of the plasma by very strong toroidal magnetic fields which constrain it to the centre of a doughnut-shaped vessel. There have been estimates of up to twenty five years for the time when fusion reactions will be commercially exploitable. Claims that the technology will be 'clean', i.e. meaning that no radioactive products are formed, apply only to the fusion stage. The tritium (T) is made by the irradiation of lithium with neutrons:

$$^{6}_{3}\mathrm{Li} + ^{1}_{0}\mathrm{n} \longrightarrow ^{3}_{1}\mathrm{H(T)} + ^{4}_{2}\mathrm{He}$$

the only reasonable source of neutrons being a fission reactor. There are neutrons produced in the process which have to be slowed down and captured. When they are captured they produce radioactive isotopes of the element that is used to capture them.

The fusion process has been feasible in an uncontrolled manner since the explosion of the first 'hydrogen-bomb' in 1952. The main explosive in such devices consists of crystalline lithium deuteride made from lithium enriched in the $^{6}\mathrm{Li}$ isotope. The advantage of using a hydride (lithium deuteride) of lithium is that it is a chemically stable crystalline solid (in the absence of water) and represents an excellent method of storing the correct mixture of nuclei which is required for the fusion reaction. The high temperature required for the nuclear fusion process to occur is produced by the initial explosion of a plutonium fission bomb. A series of nuclear reactions occurs:

$$^{2}\mathrm{H} + ^{2}\mathrm{H} \longrightarrow ^{3}\mathrm{H} + ^{1}\mathrm{H}$$

$$^{1}\mathrm{H} + ^{7}\mathrm{Li} \longrightarrow 2^{4}_{2}\mathrm{He}$$

$$^{2}\mathrm{H} + ^{2}\mathrm{H} \longrightarrow ^{3}\mathrm{He} + ^{1}\mathrm{n}$$

$$^{1}\mathrm{n} + ^{6}\mathrm{Li} \longrightarrow ^{4}\mathrm{He} + ^{3}\mathrm{H}$$

$$^{3}\mathrm{H} + ^{2}\mathrm{H} \longrightarrow ^{4}\mathrm{He} + ^{1}\mathrm{n}$$

Moisture would render thermonuclear devices useless very quickly because the hydride (deuteride) ion is unstable in the presence of water and reacts to give gaseous dihydrogen:

$$\mathrm{D}^{-} + \mathrm{H_2O} \longrightarrow \mathrm{H_2}\ (\text{or HD}) + \mathrm{OD}^{-}\ (\text{or OH}^{-})$$

3.13 SAFETY ASPECTS OF NUCLEAR POWER GENERATION

The general public is encouraged to fear radioactivity by the reporting of nuclear accidents but in the history of nuclear power generation there have been only three major accidents. The first was a serious fire at the Windscale (now re-named Sellafield) site in England in 1957, the reactor overheating and causing the graphite moderator to burn. The second incident was at Three Mile Island in the USA in 1979 when a reactor core overheated and became out of control. The third, and by far the worst, was the Chernobyl reactor melt-down which occurred in 1986. Only in the last case were there any immediate deaths, thirty one people dying from severe radiation sickness within a few weeks of the accident. It is probable that there will be many more deaths in the Chernobyl area amongst those of the population who suffered high doses of radiation from the dust settling out of the atmosphere after the accident. There was some hesitancy by the authorities in evacuating the local population, although some 135,000 people were moved from the area. It is possible that increased incidence of leukaemia and thyroid cancer will occur in the irradiated population. There is a suspicion that the children of workers in the nuclear industry and those living close to nuclear installations exhibit a greater incidence of leukaemia than does the general population. The evidence for this is very sparse because of the very low incidence of the disease but data generated by the study of the Chernobyl accident will probably allow a decision about the continuing health hazards of nuclear plants and their associated accidents. The normal background radiation which all humans suffer in one year amounts to about 2400 microsieverts (μSv). The dose in excess of the background received by the population around Chernobyl in the first year after the accident was as much as 2000 μSv (83% extra), that being received by people living in England and Wales being as much as 100 μSv (4% extra).

The safety standards of the nuclear industries should be compared with those of the other main fuel-producing industries, gas, oil and coal. All three of the alternative industries attempt to adhere to rigorous safety standards but, even by present-day standards, they pose considerably greater risks to the health of their work forces and to the general public than those working in the nuclear industry. Risk assessment has shown that attending a soccer match is 400 times more likely to end in death as suffering the existence of the nuclear power generating industry. Driving a car is 1600 times more dangerous.

Although safety standards with regard to radioactive materials are very high in Western Countries, there have been disturbing reports of highly irresponsible activities in the former Communist block countries, the worst to date being the almost incredibly unwise jettisoning of redundant nuclear submarine reactors in the seas off northern Russia.

4

Air: the Earth's atmosphere

Development of the Earth's atmosphere. Extent and Composition. Pressure. Temperature. Structure. Warming and cooling mechanisms. Pollution. Ozone layer. Greenhouse gases. The future of the atmosphere. London and Los Angeles smogs. Separation of atmospheric gases and their uses.

4.1 ORIGIN OF THE EARTH'S ATMOSPHERE

The material comprising the Earth condensed from gaseous solar matter about 4.6 billion years ago. There have been many attempts to construct feasible explanations of its subsequent development. Such theories vary from that which would have people believe that the development took six days, to the current scientific idea that, even after 4.6 billion years, the process is still far from complete. From the outset, chemical processes operated, some of which laid the basis for the development of life, the others producing its environment.

The total mass of the Earth has been estimated to be 5976 million million million (5.976×10^{21}) tonnes, of which only 3.9% is represented by the oceans. The atmosphere contributes only 0.000086% of the total mass, but was of prime importance in the development and maintenance of life. It is now and ever shall be vital to life on Earth.

Earth's atmosphere has not always had its present composition. The current theory is that the present atmosphere has developed from that which existed around 3.5 billion years ago. According to one variation of this theory, in the beginning, 4.6 billion years ago when the solar system formed, two very large planets were formed, called protoplanets I and II. Protoplanet I disintegrated and formed the planets, Venus and Mercury. At some time around 3.5 billion years ago, the age of the oldest rocks on Earth, protoplanet II disintegrated to form Mars, Earth and the Moon. The changes are summarized in Fig. 4.1. The majority of the initial atmosphere was lost in this cataclysmic event, the precursor of the present day atmosphere being formed mainly as the result of an out-gassing process, including volcanic action, from the hot newly-formed Earth.

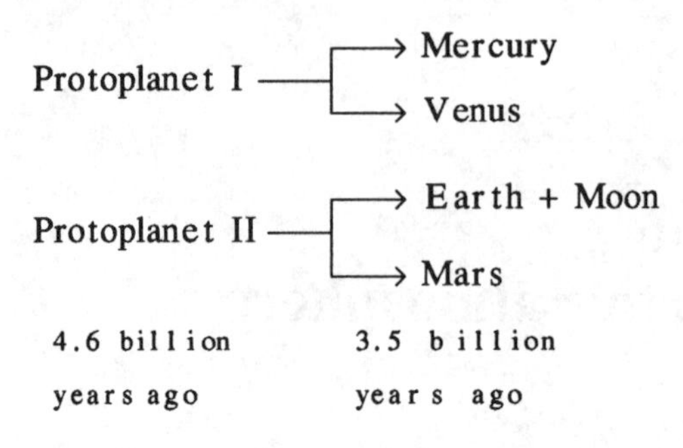

Fig. 4.1 A representation of a theory of the genesis of the inner planets

The early second atmosphere probably consisted of a mixture of water vapour, carbon dioxide, carbon monoxide, dinitrogen, hydrogen chloride, dihydrogen, and sulfur; those substances which are still emitted in volcanic eruptions. The inert gases, which form Group 18 of the periodic table of Fig. 2.1, were also present in low concentrations, the heavier ones probably being retained from the primary atmosphere of the large planet from which the new Earth was formed. Most of the dihydrogen molecules and helium atoms, being very light, had sufficiently high speeds to escape from the Earth's gravitational field. At the elevated temperature of the early atmosphere, some of the other molecules also had speeds sufficient to allow their escape. By the time the Earth was cool enough to allow water to condense to form the oceans, the atmosphere probably contained mainly dinitrogen, as it does at the present time, with carbon dioxide and water vapour as the other major constituents.

The carbon monoxide content was converted to the dioxide by reacting with hydroxyl free radicals (OH), produced photochemically from water vapour. The reaction is quite efficient, any emissions of poisonous carbon monoxide into the present atmosphere being transformed into the beneficial dioxide within fifty days.

Hydrogen chloride, which is hydrochloric acid when dissolved in water, reacted with rocks to form the soluble chlorides of sodium, potassium, magnesium and calcium which are now found in the oceans.

The sulfur reacted with iron and other metallic elements to give the sulfide minerals which are found in various regions of the Earth's crust.

A large amount of the carbon dioxide reacted with the solubilized calcium in the sea to form calcium carbonate (chalk), which under pressure became hardened to form limestone. Some of the limestone had its origins in the shells of marine organisms.

Between 3.2 and 1.8 billion years ago there was little or no dioxygen in the atmosphere and, therefore, no protective ozone layer. No life was possible on the dry land because of the intense ultra-violet radiation. Two processes were responsible for the production of dioxygen. The high energy band of intense ultra-violet radiation (UVC) from the Sun caused the photochemical decomposition of water molecules with the production of dihydrogen and dioxygen. The dihydrogen gas molecules escaped the Earth's gravity so that the dioxygen content of the atmosphere increased:

$$2H_2O \xrightarrow{UV} O_2 + \boxed{2H_2} \overset{\text{space}}{\uparrow}$$

It has been estimated that when the amount of dioxygen in the early atmosphere reached only 0.1% of its present level, the photolysis of water was prevented from occurring by the dioxygen itself preferentially absorbing the ultra-violet radiation. A more significant source of atmospheric dioxygen was life in the form of the first bacterial cells which were able to use visible light to convert carbon dioxide into carbohydrates and dioxygen; the life-producing process of photosynthesis (described in Section 3.5). It is thought that such early life forms were active around 3.2 billion years ago. The photosynthetic bacteria caused the dioxygen content of the atmosphere to increase at the expense of the carbon dioxide content until it reached a level of 1% of the present value. At such a concentration it became possible for organisms to exist which depended upon respiration for their energy requirements. Examination of the fossil record indicates that the dioxygen level of 1% of the present value was achieved around 2 billion years ago. The ozone layer had not fully developed so that life was still confined to shallow water which offered protection against the ultra-violet radiation which was incident upon the Earth's surface at the time.

Respiration reverses the photosynthetic process, converting dioxygen into carbon dioxide, but the rapidly multiplying photosynthetic organisms produced sufficient dioxygen to cause its atmospheric concentration to increase. When the level had built up to 10% of the present day value, the production of the ozone layer in the upper atmosphere was sufficiently developed to protect the Earth's surface from the more damaging portion of the ultra-violet radiation (UVC). This important stage was reached around 400 million years ago, so that life could begin to evolve on the land from that time. Early vegetation developed and contributed to the formation of the deposits of coal between 360 and 290 million years ago. Some of the oxygen was consumed in the formation of oxides in the outermost section of the Earth's crust but sufficient dioxygen built up in the atmosphere to sustain the protective ozone layer and so prepare the Earth for the subsequent development of its various extra life forms. The development of vegetation reduced the proportion of carbon dioxide and increased that of dioxygen as photosynthesis proceeded, so that the main constituents of the atmosphere became dioxygen and dinitrogen with a small proportion of carbon dioxide.

If life on Earth did not exist, with its continual replenishment of the dioxygen content of the atmosphere, ultra-violet radiation and lightning would ensure that any dioxygen would eventually be converted into nitrogen oxides. These would be washed out of the atmosphere into the oceans which would then be rich in nitrates.

Significant developments in the production of life-forms and in the composition of the atmosphere are shown in Fig. 4.2. Eukaryotic cells are more sophisticated than the prokaryotic cells of which photosynthetic bacteria are examples. Eukaryotic cells possess a 'nucleus' which contains their DNA and are the constituents of metazoans, which are multicellular creatures, mammals and human beings.

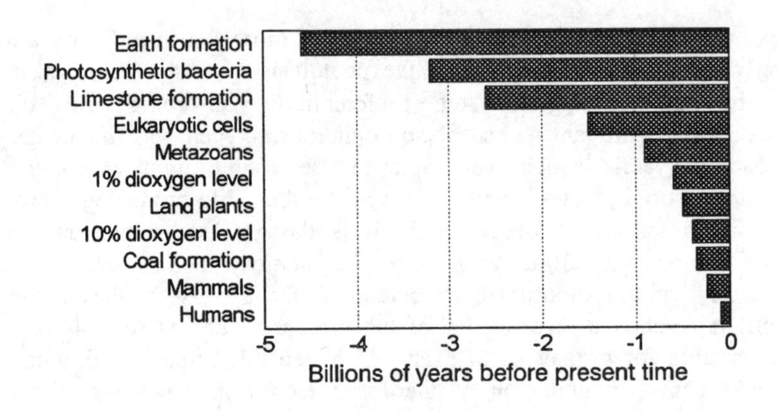

Fig. 4.2 The major developmental events in the evolution of the Earth and of life

4.2 COMPOSITION OF THE EARTH'S ATMOSPHERE

The present day composition of the atmosphere is given in Table 4.1.

Table 4.1 - The composition of the Earth's atmosphere - the main permanent components

constituent	% by volume
dinitrogen, N_2	78.083
dioxygen, O_2	20.946
argon, Ar	0.934
carbon dioxide, CO_2	0.034
others	0.003

Amongst the 0.003% of other constituents are the gases, helium (0.00052%), neon (0.0018%), krypton (0.000114%) and xenon (0.0000086%), and small amounts of methane, CH_4 (0.0002%), dihydrogen (0.00005%), and dinitrogen monoxide, N_2O (0.00005%).

Although these fractions are small, the atmosphere is sufficiently extensive to be a reasonable source of the elements, helium, neon, argon, krypton and xenon, which are extracted by the careful fractional distillation of liquefied air. The dinitrogen monoxide may be regarded as a pollutant, formed in thunderstorms. Other polluting substances such as chlorofluorocarbons - CFC's, carbon monoxide, the other oxides of nitrogen and sulfur dioxide have concentrations in the part per billion range, i.e. of the order of 0.0000001% by volume. At the average temperature of the Earth's atmosphere at sea level of 15°C, water vapour could be present to a maximum level of 0.7%, corresponding to the saturation of the air. This represents a vast quantity of water, there being fourteen million million (1.4×10^{13}) tonnes of it in the atmosphere at any time. It is not evenly distributed. In desert areas and in very cold areas, around the polar regions and at the tops of high mountains, the water vapour content of the atmosphere is minimal, whereas during the monsoon periods in tropical regions it can approach its maximum. These large variations in the amount of water vapour in the air have a very important effect upon the rate of warming and cooling of the atmosphere and are a major cause of the world's climate. In desert regions where the water content of the atmosphere is almost zero, the diurnal temperature range, i.e. between the maximum and minimum values within 24 hours, is much greater than where the atmosphere is almost saturated with water vapour. The greatest diurnal range (55°C) was observed in the northern Sahara in Tunisia, the maximum temperature was 52°C, the minimum was -3°C.

The other components of the atmosphere, being permanent gases, i.e. they are gaseous at the lowest temperature experienced on Earth, have contributions according to that of Table 4.1, which are independent of the height above sea level, up to 100 km. They are mixed mainly by the processes of convection, hot air rising to displace colder air which falls, and advection, which is the lateral movement of air forming the winds.

4.3 THE STRUCTURE OF THE EARTH'S ATMOSPHERE

Although it represents a small percentage of the Earth's mass, the amount of gas in the atmosphere of the Earth is very large. It has been estimated that the mass of the atmosphere is 5136 million million (5.136×10^{15}) tonnes. This gas is distributed around the area of the Earth which is 510 million (5.1×10^{8}) square kilometres. The depth, or height above sea level, of the atmosphere is difficult to define since the pressure decreases exponentially to virtually zero in outer space and is also dependent upon local temperature. This particular example of exponential decrease is one in which the pressure is reduced by a factor of ten for every sixteen extra kilometres of height above sea level, assuming a constant temperature. The standard atmospheric pressure at 25°C at sea level may be referred to as one atmosphere (1 atm). It is defined exactly as a pressure of 1013.25 millibars or 101.325 kilopascals (kPa). At a height of sixteen kilometres above sea level the pressure would be one tenth of 1 atm (0.1 atm), and at a height of 32 km it would be one hundredth of an atmosphere (0.01 atm). At nine times sixteen kilometres (i.e. 144 km or 89.5 miles) above sea level the pressure would be one thousand millionth of an atmosphere (0.000000001 atm) which is generally regarded, in laboratory apparatus used for gas reactions, as a very good vacuum. The height of 144 km, therefore, may be regarded as the virtual top end of the atmosphere for practical purposes. Even so, at altitudes greater than 144 km there is a sufficient pressure to show meteor trails. Meteors, small amounts of interstellar matter, heat up as they experience friction with the rarefied atmosphere and slow down, usually burning up in the process and disintegrating into dust particles. The *Aurora Borealis*, the Northern Lights, are produced by the interaction of the rarefied atmosphere with cosmic rays at distances of up to 960 km above the Earth's surface. The pressure at the top of the Earth's highest mountain, Everest, is only 28% of that at sea level, this being the reason that most climbers of the peak need to take a supply of extra oxygen with them.

4.4 EFFECT OF THE ATMOSPHERE ON THE EARTH'S SURFACE TEMPERATURE

The Earth's atmosphere is very largely responsible for the regulation of the temperature of the Earth's surface. The surface temperature of the Moon, which has no atmosphere because of its low gravitational attraction, any gaseous molecules escaping into space, varies between 120 K (-153°C) when it is not being irradiated by the Sun, i.e. the Moon's night, and 370 K (97°C) during its daytime. The Earth would suffer the same extremes of surface temperature if it did not possess an atmosphere and had a surface similar to that of the Moon, i.e. desert and rock. The reflexion properties of the various kinds of surface coverings of the Earth, together with the clouds of the atmosphere, combine to cause the reflexion of 30% of the solar radiation which is incident upon the planet. The percentage of solar radiation which is reflected by a planet is called its albedo. The albedo of the Moon is only 12%.

The Earth's surface temperature varies between the more moderate average values of 250 K (-23°C) and 310 K (37°C), some parts of the planet suffering smaller or larger extremes. Based upon these data, the average surface temperatures of the Earth and the Moon are 280 K (7°C) and 245 K (-28°C) respectively, the atmosphere and the different surface coverings benefiting the Earth to the extent of 35 degrees Celsius.

4.5 VARIATION OF TEMPERATURE WITH ALTITUDE

The temperature of the atmosphere varies in an irregular, but understandable, manner with altitude. Fig. 4.3 shows a graph of the average temperature of the atmosphere in degrees Celsius as the height above sea level increases. There are four regions where the temperature of the atmosphere is changing with height above sea level. These regions are called (i) the troposphere, (ii) the stratosphere, (iii) the mesosphere and (iv) the thermosphere. The three intervening regions, where the temperatures are reasonably constant, are called the tropopause, the stratopause and the mesopause, in order of their increasing distances above the Earth's surface. The four regions and their divisions are shown in Fig. 4.4. Everyone is aware of the properties of the troposphere which are known collectively as the weather.

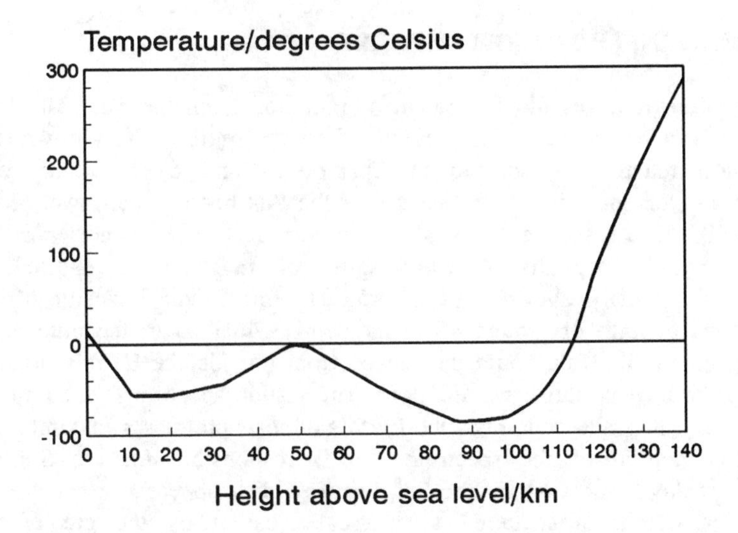

Fig. 4.3 A graph showing the variation of the average temperature of the Earth's atmosphere with the height above sea level

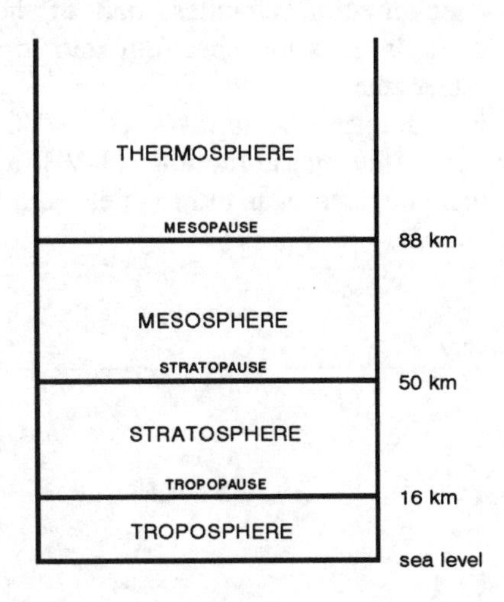

Fig. 4.4 A diagram showing the four regions of the Earth's atmosphere where the temperature varies with height above sea level and the three intermediate divisions where the temperature is reasonably constant

The thermal variations of the atmosphere may be understood in terms of three warming influences:

(i) the absorption of high energy radiation from the Sun,
(ii) the absorption of the majority of the ultra-violet radiation from the Sun, and
(iii) the absorption of infra-red radiation from the heated Earth.

4.6 WARMING IN THE THERMOSPHERE

The thermosphere receives the full range of radiation from the Sun. The interior of the Sun has a temperature in the region of fifteen million Kelvin which allows nuclear fusion reactions (discussed in Chapters 3 and 6) to occur readily. It functions as an element factory. The surface of the Sun has a much lower temperature of around 6000 K. At such a temperature the Sun acts as an almost perfect cavity radiator. A broad spectrum of wavelengths of radiation is emitted, with a distribution of intensity shown in Fig. 4.5. The maximum intensity of radiation occurs at a wavelength of around 485 nanometres which is in the blue part of the visible spectrum. If the Sun could be viewed from outside the Earth's atmosphere it would appear to have a bluish white colour. The visible spectrum, which survives the journey from the Sun and penetrates the Earth's atmosphere, has a maximum intensity in the yellow region of the spectrum and is responsible for the Sun's yellow appearance. At dawn and twilight the Sun has a reddish appearance because the blue portion of its output is scattered to a greater extent by the greater depth of atmosphere which the Sun's light has to penetrate at those times. It is even redder when there is a considerable amount of water vapour present in the atmosphere which aids the scattering of the blue end of the spectrum. It is the scattered blue end of the spectrum that causes the day-time cloudless part of the sky to be that colour. It is useful to divide the Sun's emission spectrum into three sections as shown in Fig. 4.5, these having wavelengths;

(a) below 200 nm, including the far ultra-violet - UVC, X and γ rays,

(b) 200-300 nm, which is the mid-ultra-violet UVB, and

(c) above 300 nm, including the near ultra-violet; some UVB and all the UVA, the visible region and infra-red.

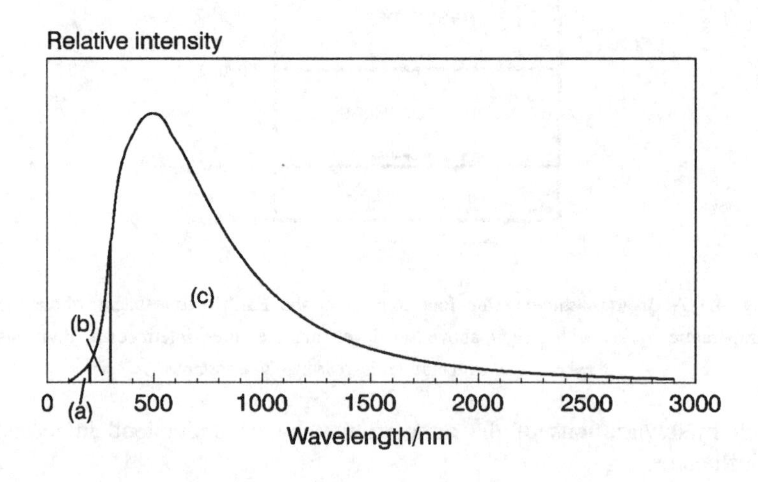

Fig. 4.5 A graph of the relative intensity of the continuous emission from the Sun as the wavelength of the radiation varies; sections (a), (b) and (c) with wavelength ranges < 200 nm, 200-300 nm and > 300 nm respectively

Section (a) of the Sun's emission spectrum causes the warming in the thermosphere. All the ionizing radiation, which includes high energy γ and X-rays, is absorbed by the atmosphere at heights above 60 km. The absorption of the radiation causes the warming. Typical processes occurring in the thermosphere are ionization of molecules to give positive ions and electrons, with some of the positive ions decomposing:

$$N_2 \xrightarrow{\gamma,X} N_2^+ + e^-$$

$$N_2^+ \longrightarrow N + N^+$$

$$O_2 \xrightarrow{\gamma,X} O_2^+ + e^-$$

$$O_2^+ \longrightarrow O + O^+$$

to give oxygen and nitrogen atoms, together with positive ions and a corresponding number of free electrons. All the particles have considerable kinetic energies, i.e. are moving with high speeds, corresponding to the high temperature of the thermosphere. The charged particles, positive ions and negative electrons, are influenced by the Earth's magnetic field and form the system sometimes called the ionosphere. The ionosphere has the property of reflecting long wavelength radio waves and allows the transmission of broadcasts such as the BBC World Service to take place over large distances. The far ultra-violet region of the spectrum (UVC) is absorbed partially by dioxygen in the thermosphere but its main contribution to atmospheric warming occurs in the lower mesospheric and stratospheric sections where the dioxygen concentration is higher.

4.7 WARMING IN THE MESOSPHERE AND STRATOSPHERE; THE OZONE LAYER

As the atmosphere becomes more concentrated below the thermosphere, a second mechanism of warming occurs. In these regions, from the mesopause, between 86-90 km above sea level, to the upper part of the stratosphere from 20-48 km high, there is sufficient dioxygen to absorb all the ultra-violet radiation with wavelengths below 200 nm (UVC) and a large fraction of section (b) of the Sun's emission spectrum (UVB and UVA). The wavelengths between 200-240 nm cause atmospheric warming as the radiation is converted into the energy of translation of the dioxygen molecules and is ultimately shared out between all the components of the atmosphere. Absorption of the wavelengths below 200 nm causes dioxygen molecules to photodissociate into oxygen atoms:

$$O_2 \xrightarrow{UV < 200\ nm} 2O$$

The original ultra-violet photon energy is converted to the kinetic energy of motion of the oxygen atoms. They move with greater speeds which is equivalent to the atmosphere having a higher temperature. Some of the oxygen atoms combine to give dioxygen.

A very important reaction of oxygen atoms occurs in the 20-40 km high region of the stratosphere. Oxygen atoms react with dioxygen to give molecules of trioxygen or ozone, O_3:

$$O + O_2 \longrightarrow O_3$$

The ozone molecules then absorb the portion of the ultra-violet radiation from the Sun which has wavelengths between 240 and 300 nm. In absorbing the ultra-violet radiation, some of the ozone molecules decompose photochemically according to the equation:

$$O_3 \xrightarrow{\text{UV 240-300 nm}} O_2 + O$$

The oxygen atoms destroy other ozone molecules by the reaction:

$$O + O_3 \longrightarrow 2O_2$$

the ozone being constantly replaced by further combinations of oxygen atoms with oxygen molecules. The rates of production and destruction of ozone are approximately equal, which allows an equilibrium concentration of the gas to exist in this part of the atmosphere. There are other reactions which lead to ozone depletion, but which also participate in the determination of the steady state concentration of the gas. Naturally present in the stratosphere are small concentrations of of water, nitrogen monoxide and nitrogen dioxide, the latter two being formed by the reaction of nitrogen atoms with oxygen. Water molecules undergo photodissociation by ultra-violet radiation with wavelengths below 200 nm according to the equation:

$$H_2O \xrightarrow{\text{UV < 200 nm}} H + OH$$

to give hydrogen atoms and hydroxyl free radicals. The hydrogen atoms can destroy ozone by the reaction:

$$H + O_3 \longrightarrow OH + O_2$$

and the hydroxyl radicals continue the ozone destruction by the reactions:

$$\rightarrow OH + O_3 \longrightarrow HO_2 + O_2$$

$$HO_2 + O_3 \longrightarrow OH + 2O_2$$

This is an example of a chain reaction. The hydroxyl radicals are regenerated by the reaction of the hydroperoxyl radicals (HO_2) with ozone and then continue the chain mechanism. The unstable free radicals, OH and HO_2, are examples of chain carriers. They participate in the propagation of the chain reaction.

The nitrogen oxides also destroy ozone by a chain reaction which is expressed by the equations:

$$\rightarrow NO + O_3 \longrightarrow NO_2 + O_2$$

$$NO_2 + O_3 \longrightarrow NO + 2O_2$$

It might appear that all the ozone should be destroyed, but an appropriate fraction of the chain carriers, i.e. the unstable highly reactive species, OH, HO_2, NO and NO_2 that propagate chains, react with each other to give relatively stable products, for example:

$$OH + HO_2 \longrightarrow H_2O + O_2$$

$$OH + NO_2 \longrightarrow HNO_3 \text{ (nitric acid)}$$

The above reactions, together with many others, occur at various specific rates and have combined to produce an equilibrium, or steady state, concentration of ozone in the stratosphere which is sufficient to protect the Earth's surface from radiation with a wavelength lower than 300 nm. The ozone is being generated at the same rate that it is being destroyed so that its concentration is constant. There are some fears that pollutants such as chlorofluorohydrocarbons (CFC's, e.g. CCl_2F_2 which is known as Freon 12; one carbon atom, two fluorine atoms) are contributing to ozone destruction and causing a depletion in the equilibrium level of stratospheric ozone. CFC's are used in refrigerators, spray cans of various kinds, e.g. perfumes and paints, and for producing articles from expanded polystyrene. Their escape into the atmosphere should not be permitted as they are very volatile, and are a source of chlorine atoms if subjected to ultra-violet radiation. Chlorine atoms react with ozone according to the equations:

$$\begin{array}{l} \rightarrow Cl + O_3 \longrightarrow ClO + O_2 \\ ClO + O_3 \longrightarrow Cl + 2O_2 \end{array}$$

and would be prevented from continuing the chain process by reaction with another atom or small molecule. The chain decomposition of ozone by chlorine atoms is considerably faster than that caused by hydroxyl radicals or nitrogen oxides, so that a small concentration of chlorine can be very destructive. The use of CFC's is being phased out by international agreement since the seriousness of the problem was recognized. Whether this course of action has been justified will emerge in future years.

There is no doubt that chlorine atom destruction of the ozone layer is occurring, but there are some doubts about the source of chlorine atoms. The doubts are based upon the extremely small concentrations of CFC's in the troposphere and the difficulty that their relatively heavy molecules would have in diffusing through the tropopause and into the upper reaches of the stratosphere in sufficient concentration to initiate the chain decomposition of ozone. The temperature of the stratosphere increases as the height above the Earth increases, as is shown in Fig. 4.3. This prevents the occurrence of mixing by convection, the only mechanism being that of diffusion, the rate of which is inversely proportional to the square root of the mass of the molecule. The rates of diffusion of dioxygen and Freon 12, which is the lightest CFC, would be in a ratio of almost 2:1. The larger Freon molecules would diffuse into the upper regions of the atmosphere even more slowly.

An alternative source of chlorine atoms is the combustion of the fuel used in the booster rockets for the Space Shuttle flights. The fuel is a mixture of ammonium perchlorate (NH_4ClO_4), much used in terrestrial fireworks, and powdered aluminium metal, bonded together with a hydrocarbon polymer. The products include hydrogen chloride (HCl), dichlorine (Cl_2), nitric oxide, carbon dioxide, water and aluminium oxide, Al_2O_3. Water is also produced from the main rocket engine which burns dihydrogen and dioxygen. The products from a single shuttle launch which are ejected into the stratosphere are given in Table 4.2.

Table 4.2 - The major stratospheric emissions from a single launch of the Space Shuttle

product	amount/tonnes
water	146
carbon dioxide	148
aluminium oxide	110
hydrogen chloride	60
dichlorine	12

There is a relatively small quantity of nitric oxide produced in each launch, amounting to about 0.3 tonnes. The hydrogen chloride and dichlorine are easily photolyzed to give the ozone destroying chain carriers, chlorine and hydrogen atoms. As each Shuttle blasts off, the substances are emitted at an altitude where they happen to be able to do maximum damage to the ozone layer. They do not have to diffuse from sea level, 20-40 km below, as do the CFC's.

Recent evidence suggests that the CFC's tend to condense out of the atmosphere in the two very cold polar regions of the Earth and it is likely that some condensation also occurs at the tops of icy mountains. Both sources of atmospheric pollution are currently subjects of vigorous study, the actual cause of any observable stratospheric ozone diminution having yet to be determined. Ozone diminution is observed in the polar regions, but these are exactly the regions where the ozone production is least because they receive much less intense irradiation from the Sun.

4.8 COOLING OF THE THERMOSPHERE AND THE STRATOSPHERE

The thermosphere and stratosphere have average temperatures as indicated in Fig. 4.3. Their warming mechanisms are described in some detail in Sections 4.6 and 4.7. Both regions of the atmosphere maintain their average temperatures by cooling while they are not directly in the Sun's path. The processes of ionization and molecular dissociation are reversed with the emission of radiation which escapes into space, i.e. that fraction which is directed away from the Earth. Any molecules of carbon dioxide which are in a state of vibrational excitation lose their energy by emitting radiation with characteristic wavelengths appropriate to the vibrations. Details of the rules governing the emission of radiation by molecules are described in the next section. A sufficient fraction of the emitted radiation escapes into space to offset the warming which occurs when the atmospheric regions are receiving solar radiation.

4.9 WARMING AND COOLING IN THE TROPOSPHERE

The Sun's radiation which penetrates the upper parts of the atmosphere contains a small fraction in the ultra-violet region (300-400 nm, UVB and UVA), the visible light upon which all life depends, and a considerable amount of infra-red radiation, these being represented in Fig. 4.5 as section (c). Some of the ultra-violet and visible radiation in section (c) is reflected away from the Earth, but the majority is absorbed by the Earth's surface, resulting in the warming of the surface.

An additional contribution to the warming of the surface is heat conducted from the interior of the Earth. The temperature of the Earth's core is around 4000 K, but rocks are very good thermal insulators so that the rate of heat loss to the surface is very low. The thermal gradient of the solid earth, near the surface, is a rise of 30°C per kilometre of extra depth, a fact which is well known to operators of deep mines who have to install expensive cooling facilities. The average figure for the warming of the surface by heating from the core is 0.08 watts per square metre which, compared with the estimated 155 watts per square metre arising from solar radiation, is almost negligible (0.05%). Under local circumstances of volcanic eruption, the heating of the surface from within is far from being negligible.

The warmed Earth's surface emits radiation, particularly during the night as it cools down. The average temperature of the Earth's surface has been estimated to be 15°C or 288 K and, as such, acts as a broad continuous-spectrum radiator at that temperature. The wavelength of maximum intensity emission of a perfect cavity radiator at 288 K is 10 micrometres (μm) or 10000 nanometres, in the infra-red region as shown in Fig. 4.6. The radiation is directed generally outwards from the surface and is called terrestrial radiation.

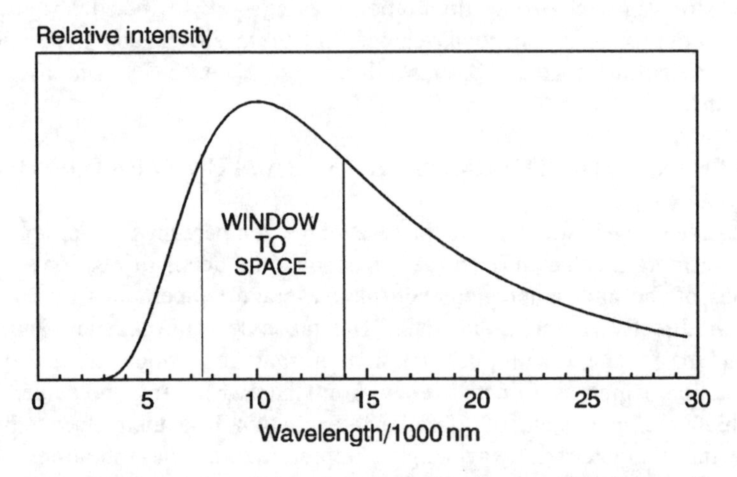

Fig. 4.6 The variation of the relative intensity of terrestrial radiation with wavelength; the infra-red window from 7.5-14 μm is shown; this allows the Earth's radiation to escape into space

The constituents of the atmosphere which absorb terrestrial radiation are carbon dioxide and water vapour. The main constituents of the atmosphere, dinitrogen and dioxygen, do not absorb any infra-red radiation. The rules which govern the interaction of infra-red radiation and matter indicate that highly symmetrical molecules such as dioxygen and dinitrogen are inactive. A molecular vibration can only be activated if the vibration causes the polarity of the molecule to vary. This is not possible for O_2 and N_2. The fundamental vibrational modes of the water and carbon dioxide molecules which are infra-red active are shown in Fig. 4.7, the word fundamental being used in the same manner as in the description of musical notes. The modes for water are the two stretching and the bending vibrations which are responsible for absorptions at wavelengths of 2.5-3.0 μm, covering the two stretches, and 5-7.5 μm respectively. The modes for carbon dioxide are the anti-symmetrical stretching and the bending vibrations which are responsible for absorptions at 4.0-4.5 μm and 14-16 μm respectively. These active vibrations are associated with changing polarities of the molecules, both molecules being polar because they each contain two different atoms. Overtone and combination bands, i.e. like chords in music, occur weakly at lower wavelengths. When the molecules absorb the infra-red radiation they become vibrationally excited, but very soon lose the vibrational energy when they collide with any other molecules, e.g. O_2 or N_2. The two molecules participating in the collision share out the vibrational energy and both have a higher speed in consequence. This is equivalent to stating that the temperature increases, and is the mechanism of warming adopted by the troposphere with regard to the incoming radiation.

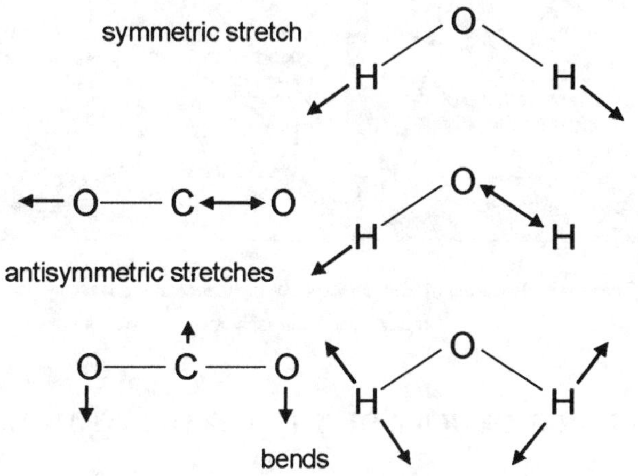

Fig. 4.7 The fundamental vibrational modes of water and carbon dioxide which are active in absorbing infra-red radiation

The terrestrial radiation is absorbed by water vapour and carbon dioxide in the atmosphere at the characteristic wavelengths mentioned above. The radiation in the regions below 7.5 μm and above 14 μm is totally absorbed in a layer which, at sea-level, is about thirty metres deep. This means that any further increase in the carbon dioxide concentration will have no extra effect. The other radiation, mainly in the range between 7.5-14 μm escapes into space, unless it is absorbed or reflected by clouds or dust particles, and is the cooling mechanism of the troposphere. This range represents an infra-red 'window' to outer space and is indicated on the graph in Fig. 4.6. The radiation at wavelengths above 14 μm is completely absorbed by water molecules which increase their rotational energy by the process. The rotational energy is eventually converted to kinetic energy causing an increase in temperature. Fig. 4.8 summarizes the main features of the absorption of solar and terrestrial radiation by the Earth's atmosphere.

The account of atmospheric warming given in this chapter is a simplification of an extremely complicated subject. Care has been taken to avoid over-simplification, the main factors affecting the temperature of the atmosphere being treated in a substantially proper manner.

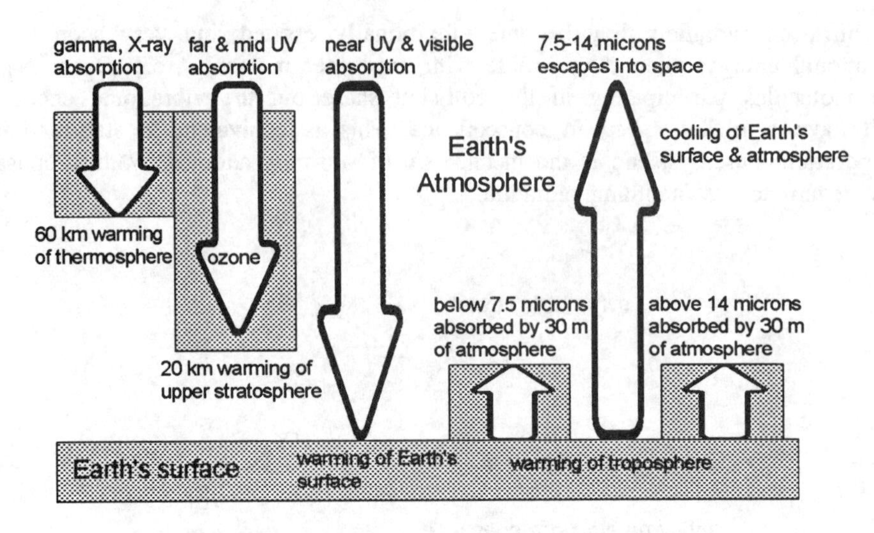

Fig. 4.8 The main features of the absorption of solar and terrestrial radiation by the Earth's atmosphere

4.10 ALLEGED ENHANCEMENT OF THE EARTH'S TEMPERATURE

The absorption of terrestrial radiation by water vapour and carbon dioxide is regarded as being analogous to the solar heating that takes place in a greenhouse. In that case, solar radiation passes through the glass and warms the surfaces inside. The surfaces then re-emit the absorbed radiation in the infra-red region which is then absorbed by the water vapour and carbon dioxide of the atmosphere inside the greenhouse, or is absorbed or reflected by the glass which constitutes the majority of the greenhouse. The glass used for glazing greenhouses, i.e. soda glass as described in Chapter 8, is opaque to infra-red radiation, the absorption and reflection of which contributes considerably to the enhanced heating which occurs in greenhouses. Greenhouses cool down by the conduction of heat through the glass which is a slow process.

Water vapour, carbon dioxide and other gases emitted into the atmosphere are fashionably known as 'greenhouse' gases. The role of water vapour is largely ignored because the amount of it in the atmosphere cannot be controlled, but water vapour is the most important contributor to atmospheric warming by the absorption of terrestrial radiation. Carbon dioxide emissions from the burning of wood and fossil fuels, together with those emanating from cement manufacture (about 3% of the total), are currently blamed by some environmentally interested groups for an alleged increase in the average temperature of the Earth's atmosphere. There is no doubt that the concentration of atmospheric carbon dioxide has risen over the course of the twentieth century, particularly in the second half. Fig. 4.9 shows a plot of the carbon dioxide concentration in the atmosphere for the years between 1958 and 1989 which were recorded at the Mauna Loa observatory in Hawaii. The annual cycles are consistent with the main growing season in the Northern Hemisphere, where most of the world's vegetation exists, tending to their minima towards the end of the Northern Summer around September and October when the cultivated plants have been harvested and the general vegetation is growing more slowly. The maxima correspond to the end of the Northern Spring in May when less heating is required and there is less burning of wood and fossil fuels, and before the next growing season has commenced. It was considered that the air contained 0.028% of carbon dioxide at the beginning of the century, the latest figure of 0.035% representing an increase of 25%. It is assumed that there is a fairly rapid mixing of the constituents of the atmosphere and that the measurements at Mauna Loa are representative of the whole troposphere.

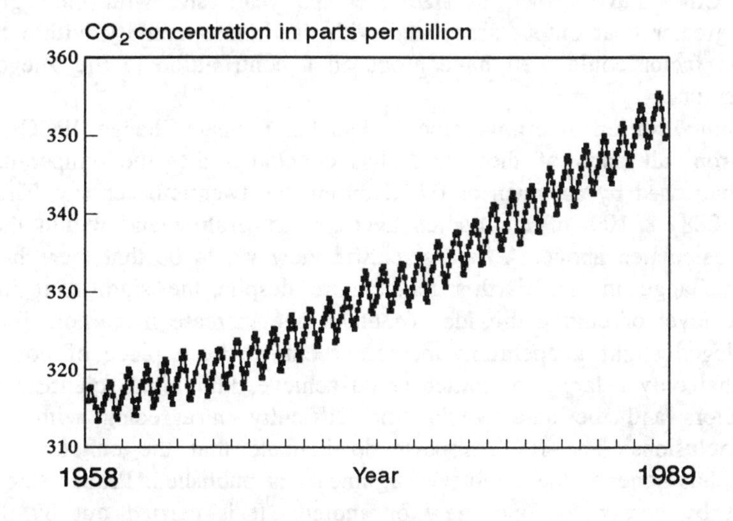

Fig. 4.9 The variation of carbon dioxide concentration in the atmosphere for the years between 1958 and 1989; recorded at the Mauna Loa observatory in Hawaii

There are some difficulties associated with the temperature record over the century. These are (i) the doubtful accuracy of the thermometers used and the accuracy to which these are read in the monitoring stations around the world, (ii) the use of computed average temperatures and (iii) the geographical distribution of the monitoring stations. Thermometers used for recording atmospheric data are usually the conventional mercury-in-glass type, perfected by Fahrenheit, now graduated in degrees Celsius. The recorded temperatures are usually quoted to the nearest degree Celsius, although the thermometers used are capable of being read with reasonable accuracy to a tenth of a degree. All thermometers of this type suffer from the impossibility of supplying continuously recordable data so that daily average temperatures are not particularly accurate. Annual mean temperatures computed from such daily figures suffer from the same deficiencies. The daily average temperature is normally taken as the average of the maximum and minimum values and takes no account of the variation between them. The average temperature over a twenty-four hour period would be more reliable if hourly readings were recorded. Even greater accuracy would be afforded by taking temperature readings every minute, such a procedure being rather hard on the operators. The thermometers which would allow proper average daily temperatures to be recorded accurately are the electronic types with integration facilities. Otherwise the quoted averages are at least only as accurate as one temperature reading which could be in error by as much as one degree in 288 (the supposed average temperature of the Earth's lower atmosphere), i.e. 1/288 x 100 = 0.35%. The geographic distribution of measuring stations around the world has not been constant over the century as newer stations have been opened. Nor is their distribution regular with respect to the Earth's surface. Some measurements are taken by ships on the water which covers 71% of the Earth's surface, but the majority of stations are on land and are mainly found in the cities. Cities have grown in size over the years and with their greater size there is a greater heat output associated with the human activity within them. This urbanization factor could well have produced a contribution to the alleged heating over the century.

The United Nations Intergovernmental Panel on Climate Change (IPCC), a group of scientists from all parts of the world, has concluded that the temperature of the Earth has increased by as much as 0.8°C during the twentieth century. This is 0.3%, i.e. (0.8 ÷ 288) x 100, of the quoted average temperature and within the possible error limit calculated above. A hard scientific view would be that there has been no discernible change in the Earth's temperature despite the significant increase of 25% in the level of carbon dioxide. To blame the increase in carbon dioxide level for this alleged slight temperature increase seems to be a piece of poor scientific judgment that only a large committee could achieve. Individual scientists looking at all the factors and the data would find difficulty in agreeing with the panel's (IPCC) conclusions. The IPCC reports do indicate that the conclusions are not unanimous, but none of the doubter's arguments is published. Proper science is not carried out by voting for one view or another. It is carried out by the serious scrutiny of observations by the scientist and his or her peers, and by the further collection of data to test any preliminary conclusions. Only when doubter's arguments have been disposed of by hard experimental observations that pass the peer review test should the conclusions be available for the basis of future actions.

There are other factors which may have influenced the temperature of the Earth over the 20th. century. One is the occasional catastrophic volcanic eruption, e.g. such as those which occurred at Mount St.Helens, USA, in 1980 and Mount Pinatubo, Philippines, in 1990, which sends vast quantities of gas and dust into the upper atmosphere. The dust clouds take from between six months to five years to settle out and can significantly alter the amount of radiation reaching the Earth's surface and the amount which escapes. The Krakatoa volcanic explosion in 1883 threw sufficient dust into the upper atmosphere to have a serious effect upon the world's climate for several years.

The other method of placing dust into the upper atmosphere is to test H-bombs. Between 1945 and 1980 there were 423 such tests in the atmosphere, the greatest number being detonated in the years between 1956-1962 inclusive, when 286 were exploded. Any tests after 1980 have been conducted underground and are not relevant to the argument, no atmospheric pollution being produced by them. Dust in the atmosphere is usually removed as water falls to Earth as rain, hail or snow.

Another natural factor which may influence the Earth's temperature is the variation in the so-called solar constant, the amount of radiation reaching the Earth from the Sun. The amount is reasonably constant but does vary, particularly with the number of sun-spots on the Sun, to an extent of ±2% over an eleven to thirteen year cycle. Such a variation is larger than the alleged 0.3% increase in the Earth's temperature over the last century. Recently, it has been shown by John Butler, of the Armagh Observatory, that there is a highly significant correlation between the length of sun-spot cycles and the average temperature at Armagh, the shorter the length of a cycle, the greater being the average temperature. He makes a very plausible case for the conclusion that, between 1795 and the present day, any variations in the average temperature at Armagh have been due to variations in the Sun's output, and have had no connexion with changes in the composition of the atmosphere over that time. The Armagh thermometers have been re-calibrated and the temperature record is not subject to the effect of urbanization as the size of Armagh has changed little since 1795.

Any emission into the atmosphere of molecules which have infra-red absorptions in the 7-14 μm range, the main 'window' through which emission of radiation causes the Earth to cool down, could be a possible factor influencing the cooling mechanism. So-called greenhouse gases with such properties include sulfur dioxide (the precursor of acidic rain), methane and the problematic CFC's. Sulfur dioxide absorbs at around 7.4 μm and 8.7 μm and would contribute to atmospheric warming but for its solubility in water to give sulfur(IV) acid, H_2SO_3. This is easily oxidized to sulfur(VI) acid (sulfuric acid), H_2SO_4, which is removed from the atmosphere periodically as acidic rainwater. Methane absorbs at 7.7 μm, and CFC's possess several absorptions in the 7-14 μm region, the gases being enhancers of global warming if present in sufficient quantities. Methane is not a serious threat to the environment since it undergoes decomposition, by oxidation to carbon dioxide and water via ultra-violet irradiation or in lightning, reasonably quickly. On average, a molecule of methane is oxidized to carbon dioxide within a period of four years of its emission into the atmosphere. The CFC molecules are very stable and do pose a threat if they do not condense substantially in the arctic and antarctic regions. Their manufacture is being phased out, mainly because of their threat to the ozone layer. Their replacements are supposed to pose a smaller threat to the ozone layer but still would be greenhouse gases in that they absorb in the vital window region. Their condensation in the colder regions would largely remove such a threat.

The current debate about global warming started with a paper Sawyer (*Nature* **239**, 23, 1972) which is an excellent and properly scientifically cautious account of the subject. Sugden's Royal Society Study Group Report, *Pollution in the Atmosphere*, 1978, states that '*an increase in atmospheric CO_2 has been thought by some to expose the planet to the dangers of a greenhouse effect*' and goes on to state that '*Reliable estimates are extremely difficult to make particularly as the resultant effect is small - much less than the seasonal variation due to the growth of vegetation - and is ameliorated by other factors which tend to decrease the average temperature. The prudent course is to conserve resources, minimize wasteful practices and collect reliable data on the magnitude of the various carbon-containing reservoirs and the rates of transfer between them*'. Since 1978, no further progress has been made in understanding the effects of greater carbon dioxide emissions, but there has been the report of the IPCC in 1990 which has received much publicity and has caused governments of some countries of the world to consider emission limitation. The IPCC has set up a computer model of climate change which has produced some alarming forecasts of how the Earth's temperature might change over the next 110 years. The panel's 'best' predictions are added to the observations of average temperatures from 1900 to 1992 in the graph shown in Fig. 4.10. The most pessimistic prediction is that the Earth's temperature will rise by 1°C by the year 2050 and that will only take place if emissions of carbon dioxide are subject to drastic limitations. There are other brave attempts at modelling the climate which all tend to predict severe changes associated with a temperature increase of up to 5°C by the year 2100. About fifteen years ago some of the same workers were expressing serious concern about a predicted future *ice age*.

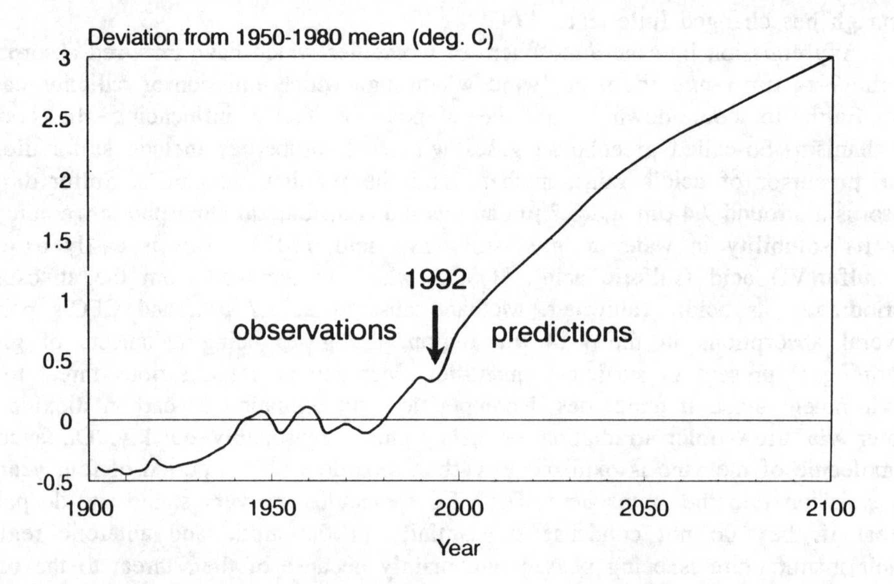

Fig. 4.10 A graph showing observations of deviations of the Earth's temperature from the 1950-1980 mean from 1900-1992 and the IPCC 'best' predictions of the temperature deviations until the year 2100

Computer models of relatively simple and well defined systems, e.g. a chemical reaction taking place in a small flask, often fail and it is inevitable that the modelling of a system as large and unwieldy as the Earth's atmosphere will produce amazing and wayward results. The IPCC's computer program cannot reproduce the last century's climate changes which bodes ill for its claim to able to predict climate changes in the next century. The prediction of tomorrow's weather is a difficult enough task. The prediction of the next day's weather is reasonably good at the present time, but five-day forecasts are still fairly poor. To consider the state of the world's climate in 110 years from now is beyond accurate scientific judgment at present. Serious studies of the effects of atmospheric composition upon climate should continue and undoubtedly will.

The main problems are those concerned with the understanding of the effects upon the warming and cooling of the atmosphere by dust particles and aerosols, i.e water droplets surrounding small particles of matter. Currently, there is little understanding of these problems and this allows an escape route for the believers in global warming. If global warming does not happen, they can claim that any warming is being off-set by the cooling effects of dust and aerosols. It is very difficult to decide whether a continuing unchanged climate is due to there being no influence from human activities or whether the effects of such activities are cancelling each other.

A quotation from Preston Cloud Jn. and Aharon Gibor's contribution to the textbook, *Understanding The Earth*, summarizes the proper scientific attitude to considerations of the future of the Earth's atmosphere. '*He who is willing to say what the final effects of such processes (man-made emissions) will be is wiser or braver than we are. Oxygen in the atmosphere might be reduced several percent below the present level without adverse effects. A modest increase in the carbon dioxide level might enhance plant growth and lead to a corresponding increase in the amount of oxygen. Perhaps we should be more alarmed about a possible decrease of atmospheric carbon dioxide, on which all forms of life ultimately depend. The net effect of burning fossil fuels may in the long run be nothing more than a slight increase in the amount of limestone deposited.*'

It is essential that all possible means should be adopted to reduce harmful emissions. The mechanisms of action of pollutants should be studied to aid in the minimization of their effects, if any. In particular, the resources of the Earth should be used in as prudent a manner as possible. The temperature of the Earth is being properly monitored by satellite measurements and over the last 10-12 years has not changed appreciably. What has been observed are variations of as much as ±0.5°C over a period of only two weeks! The monitoring equipment is accurate to ±0.01°C. Such monitoring will be continued and as more accurate data are produced will lead to a greater understanding of the atmosphere and may lead to a reasonable projection of its future. Although the mass of carbon dioxide in the atmosphere is very large and increasing at the present time, there is five times as much still in existence in the form of fossil fuels; natural gas, petroleum and coal, fifty-two times as much dissolved in the oceans, and around thirty-six thousand times as much locked away in the carbonate rocks, e.g. marble, limestone and dolomite.

4.11 SMOG

Smog, which is local low-level atmospheric pollution that can be detrimental to health, is produced in urban areas under particular conditions. There are two types of smog, typified by those that have occurred in London, England, and those that still occur in Los Angeles.

London smog is largely a thing of the past since the passing of the Clean Air Acts and amendments from 1963 onwards. London smogs were produced by industrial and domestic coal fires which release soot, i.e. particles of carbon, and sulfur dioxide into the atmosphere. The conditions for the production of smog are high levels of pollution coupled with high pressure. The high atmospheric pressure is usually associated with cloudless skies that produce cool nights followed by days of sunshine. The cool nights produce fog, i.e. an aerosol of water droplets, which is intensified by the presence of particles of soot that act as nucleation sites for condensation. If the aerosol is produced with suitable depth, the air above it is warmed by the sunlight which cannot penetrate the smog. This causes a temperature inversion in which the smog cannot disperse by convection and the further burning of fuels throughout the day causes the smog to become even thicker. The sulfurous smog is only blown away or washed out by rain when the next bank of cloud comes over the Atlantic ocean to solve the problem.

The smogs of Los Angeles, and to a lesser extent of London in Summer, are photochemically sustained. Smog formation depends upon the production of substantial vehicle-exhaust pollution and high pressure conditions with intense sunlight. The day-time temperature is required to be above 18°C for the photochemically induced reactions to be effective. The cooler cloudless nights encourage the formation of fog. The fog is intensified by the vehicle-exhaust pollution, and temperature reversal occurs, the atmosphere above the smog being at a higher temperature than the smog itself. Again, as with sulfurous conditions, convective removal of the smog is prevented.

If the temperature of the smog is above 18°C, photochemical action becomes very effective. Vehicle exhaust systems produce some nitric oxide:

$$N_2 + O_2 \longrightarrow 2NO$$

Nitric oxide is easily oxidized to nitrogen dioxide by reaction with dioxygen:

$$2NO + O_2 \longrightarrow 2NO_2$$

Nitrogen dioxide absorbs radiation with wavelengths lower than 400 nm (ultra-violet) and undergoes photodissociation to give oxygen atoms:

$$NO_2 \xrightarrow{UV} NO + O$$

The oxygen atoms then react with dioxygen molecules to produce ozone:

$$O + O_2 \longrightarrow O_3$$

the reaction being reversed photochemically to some extent. Both oxygen atoms and ozone are powerful oxidants and react with the hydrocarbon content of the polluted atmosphere to produce a range of organic substances which contribute to the deleterious effects of photochemical smog. Nitrogen oxides also participate in the chemistry Possibly the worst compound produced is that known as peroxoacetylnitrate, or ethanoyl nitrate, the formula of which is:

$$CH_3\underset{\displaystyle O}{\underset{\|}{C}}\text{-O-O-}NO_2$$

This, and other toxic irritant organic compounds, are responsible for the health hazards of photochemical smog, with higher than normal ozone levels also contributing. Los Angeles smogs are longer lasting than elsewhere because of the general stability of the local climate and the high levels of vehicle-exhaust pollution that exist.

4.12 OXYGEN

Gaseous oxygen is essential for life, providing the oxidative power for the human organism to derive energy from its intake of nutrient solids and liquids, i.e. food and drink. The energy equivalents of various foods are discussed in Chapter 9. Liquid oxygen is manufactured on a vast scale, something like 100 million tonnes per year being used for a variety of purposes. It is a very pale blue liquid, there being a very weak absorption band towards the red end of its visible spectrum, centred at 758 nm. It is produced by the fractional distillation of liquefied air. Before liquefaction the water and carbon dioxide in the air are removed by cooling and by absorption into concentrated sodium hydroxide solution respectively. The boiling points of helium (-268.9°C, 4.2 K), neon (-246.1°C, 27 K), nitrogen (-195.8°C, 67.4 K) and argon (-185.9°C, 87.3 K) being below that of oxygen (-183°C, 90 K) allows them to be boiled off from liquid air, leaving liquid oxygen with a slight content of krypton (b.p. -153.2°C, 120 K) and xenon (b.p. -108°C, 165 K). Further fractionation operations are used to separate all the constituents.

The main use for liquid oxygen is in the steel industry in which the classical Bessemer converters, which use blasts of air to oxidize some of the carbon content of the liquid iron, are being replaced by oxygen lance methods. The liquid oxygen is manufactured on the site of the steel works and injected into the furnace as the gas. The burning of diesel fuel with oxygen is replacing the conventional coke/air system of the blast furnace in the manufacture of metallic iron. Other uses for gaseous oxygen are in the production of high temperature flames (hydrogen/oxygen, acetylene/oxygen) for use in welding and other metallurgical operations. Oxygen is used chemically for the production of ethylene oxide, C_2H_4O, and propylene oxide, C_3H_6O, for the respective manufacture of polyethylene (polythene) and polypropylene and in the production of pure titanium dioxide from its chloride. As the liquid it is used to react with liquid hydrogen in the main rocket engines of the Space Shuttle. The gas is used for medical purposes and in the manufacture of hydrogen peroxide.

4.13 NITROGEN

Nitrogen is used in the synthesis of ammonia which is the basis of the nitrogen fertilizer industry. It is used to provide inert atmospheres for oxygen-sensitive reactions and as the liquid as the coolant for low temperature work, liquid air being less safe if any oxidizable materials are being used.

4.14 THE INERT GASES

The inert gases of Group 18 of the periodic table, helium, neon, argon, krypton and xenon, are used as fillers for discharge tubes for fluorescent lighting and more specialized equipment, xenon discharge tubes being a good source of far ultra-violet radiation, argon being used for filling ordinary tungsten-filament light bulbs and for inert atmospheres for specialized welding operations. Liquid helium is a safe filler for balloons, there being no reaction with dioxygen unlike dihydrogen with which the doomed Hindenburg air-ship was filled, and as the liquid (boiling at 4.2 K, -268.9°C) provides the necessary cooling for superconducting magnets used in nuclear magnetic resonance (nmr) spectrometers and scanners.

4.15 CARBON DIOXIDE

Although some carbon dioxide is produced in the liquefaction of air, it is mainly manufactured by the oxidation of methane, extracted from flue gases or obtained as the product of fermentation of sugars. Such manufacture is carried out on a large scale, the annual world production being in the region of one hundred million tonnes.

A quarter of the production is used in the carbonation of beverages. Other uses of the compound are in the refrigeration of food stuffs, the provision of inert atmospheres and in fire extinguishers. In liquid form it is used as a solvent which allows the extraction of caffeine from coffee beans before they are subjected to extraction by hot water to provide the non-stimulant de-caffeinated drink preferred by some people to the real thing. A cup of tea contains about twice as much caffeine as a cup of coffee. Carbon dioxide is used with ammonia in the manufacture of urea $CO(NH_2)_2$, which is used as the basis of urea-formaldehyde plastics and resins, and as a nitrogenous fertilizer.

5

The Earth's major metals

Metals relevant to daily life. Metals essential to life. Metallic elements used in their metallic state. Iron. Copper. Silver. Gold. Tin. Lead. Zinc. Nickel. Chromium. Aluminium. Metal Reserves.

5.1 INTRODUCTION

Metals became of interest to humans sometime towards the end of the stone age; about ten thousand years ago. Copper was the first metal to be used because it existed naturally as the element. It was reasonably soft and could be fashioned easily into objects and vessels. The metal was too soft to be used for tool-making but the discovery of bronze, an alloy of copper with tin, allowed useful tools to be made. Inadvertently, early bronze contained small proportions of zinc and lead which contributed to its hardness. Together, copper and bronze formed the mainstay of early civilizations. Our present culture would be very significantly poorer without copper and the other metals dealt with in this chapter. Although elemental iron, deposited on the Earth in meteorites, was used as early as 2500 B.C., it only became widely used from around 1500 B.C. Stonehenge was assembled about the same time. Iron came into use later than copper because of the difficulties in smelting a metal which has a relatively high melting point; copper melts at 1083°C, iron at 1537°C.

Of the 109 known elements, 87 of them are classed as metals; they exhibit high electrical conductivities in the solid and liquid states. They are ductile and malleable, i.e. they can be drawn out to make wire and beaten into sheets. Their clean surfaces have a characteristic sheen. Some of the metals have been known since early times, i.e. copper, silver, gold, iron, tin, antimony, mercury and lead, and most of these are relevant to our daily existence. Other important metallic elements are zinc, which was acknowledged as an element before 1500, but known as a constituent of brass as early as 20 B.C., nickel, which was discovered by Cronstedt in Stockholm in 1751, chromium, discovered by Vauquelin in Paris in 1780, and aluminium, discovered by Oersted in Copenhagen in 1825.

Magnesium, discovered by Joseph Black in Scotland in 1755 and isolated as the element by Humphrey Davy in London in 1808, and the elements sodium, potassium and calcium, which were discovered and isolated by Davy in 1807-1808, are of considerable importance to the constitution of living cells and to their functions. All four are metals, but their importance lies with their compounds and aqueous solutions.

The metals chosen for the subject matter of this chapter are those that are actually used in their elemental metallic forms; iron, copper, silver, gold, tin, lead, zinc, nickel, chromium and aluminium. From an industrial point of view they fall into three classes. Copper, aluminium, zinc, tin and lead are called major metals. Iron, chromium and nickel are steel industry metals. Gold and silver are precious metals. Other classifications would describe some as base metals, those which tarnish or oxidize easily, e.g. iron, and others as noble metals, those which do not tarnish or oxidize, e.g. gold. A now out-dated class consists of the coinage metals, copper, silver and gold which form Group 11 of the periodic table (Fig. 2.1). Of these three metals, only copper is used in everyday coinage, its price barely justifying its use with regard to the value of the coins which contain the metal as a major component (copper coloured coins). The 'silver' coloured coins in current use are fashioned from a copper/nickel alloy. Silver and gold are still used to produce limited editions of coins for commemorative purposes.

A summary of the world production in 1990 of the metals considered in this chapter is given in Table 5.1 together with the values (in US$ billions) of the metals produced.

Table 5.1 - The world production of metals dealt with in this chapter and their values (1990 data)

metal	world production in thousand tonnes	value in billion US $
iron	542700	18
copper	9026	25
silver	15	2
gold	2	24
tin	210	1.2
lead	5674	3.2
zinc	7300	10.8
nickel	873	7
chromium	3800	0.7
aluminium	18000	24

Magnesium, sodium, potassium and calcium are more important in their combined states and have an important biological role in the composition and functioning of living cells. They are dealt with as essential elements in Chapter 9. Iron, copper, chromium and zinc are discussed in this chapter. Although they are used mainly in their metallic forms, they also have important biological roles, all four elements being essential for life.

5.2 IRON

Iron was known to ancient civilizations and is very abundant, accounting for about 6.2% of the Earth's crust. It is the fourth most abundant element and the second most abundant metal. The Earth's core is thought to be a mixture of iron and nickel and with a diameter of around 3450 km, i.e. 2900 km below the Earth's surface, represents 15.8% of the volume of the planet. It is mainly responsible for the magnetic properties of the Earth, both metals being capable of being magnetized in their solid states.

Iron ore consists of the oxides hematite Fe_2O_3, magnetite Fe_3O_4, and limonite which is a hydrated hematite, and the carbonate mineral, siderite $FeCO_3$. A vast amount of iron is present in the Earth's crust in the form of iron pyrites, FeS_2. The mineral is not used for iron production to any appreciable extent because of problems associated with sulfur dioxide which is also produced. It is uneconomic to remove sulfur dioxide which would otherwise be emitted into the atmosphere to cause acidic rain (discussed in Chapter 7). It is, however, the source of much of the world's production of sulfuric acid.

The world production of iron ore in 1990 amounted to 542.7 million tonnes of contained metal with a value of $M 18000. The main producers of iron ore, rather than the metal itself, are shown in Fig. 5.1.

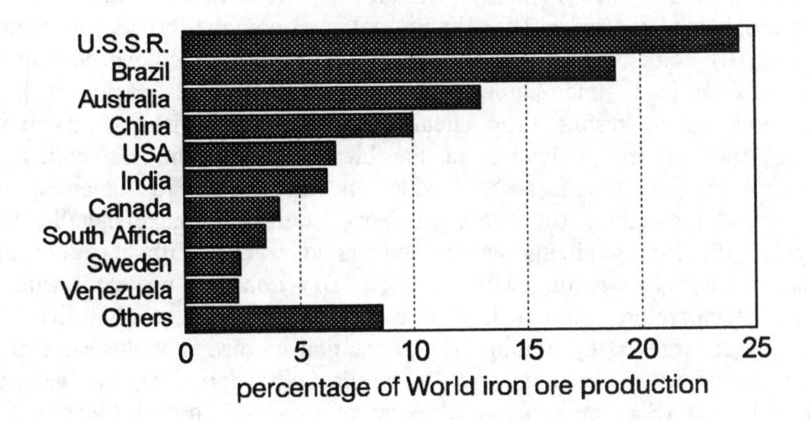

Fig. 5.1 The distribution of World iron ore production in 1990

Iron metal is produced from its ores by reducing the oxides with carbon monoxide in a blast furnace. The carbon monoxide is produced within the furnace by the partial oxidation of carbon in the form of coke. The chemistry is explained in Chapter 2. Various additives are included in the furnace in order to produce steels with specific properties.

The production of iron metal in 1990 amounted to a world total of some 692 million tonnes indicating that considerable use was made of iron ore produced in previous years. The distribution of world production of the metal in 1990 was very similar to that of the production of iron ore, the ore producing countries all having the capability of producing iron metal. The vast majority of iron ore (98%) was processed in blast furnaces in the production of various steels. The main western users were the automobile/motor vehicle manufacturers; the European Community Countries, Japan and the USA, accounting for 30.1% of the annual production (12.1%, 11.4% and 7% respectively). The U.K. in 1990 imported 14.7 million tonnes of iron ore with a metal content of 9.1 million tonnes from Australia (23%), Canada (18%), New Zealand (16%), Norway (10%) and other countries (33%). These imports represented 82% of the total consumption of iron in the U.K., the total representing only two percent of world consumption.

The domestic requirements for iron in the form of either iron, steel or stainless steel are in such products as nails, screws, bolts, steel wool, tools for wood and metal work, reinforcement rods and nets for concrete structures, cars/automobiles, washing machines and tumble driers, dishwashers, ovens, cookers and cooking utensils, refrigerators and cutlery. What would the quality of life be without iron?

The chemistry of iron in its compounds is described in terms of its two main oxidation states, now called iron(II) and iron(III), known historically as ferrous and ferric iron, respectively. The iron atom fairly easily loses two electrons in producing iron(II) compounds. The loss of a further electron is required to produce iron(III) compounds. The colours typical of the two oxidation states of iron are yellow/green(II) and red/brown(III). The red coloured common bricks owe their colour to the iron(III) content. Stock bricks which are yellow in colour contain iron in the (II) state. It is common knowledge that metallic iron is easily oxidized, the process being called rusting. The chemistry of rusting of iron is essentially the reverse of that of its production in the blast furnace. The elementary iron is oxidized to iron(III) oxide, Fe_2O_3, which is the brown rust-coloured material produced by the reaction. The two ingredients required to make metallic iron rust are oxygen gas (the oxidizing agent) and liquid water. Without either of these participating substances rusting will not occur. Dry iron will not react with oxygen at ambient temperatures, nor will iron rust in water that has had its dissolved oxygen content removed by boiling. The prevention of rusting is the basis of several industries including the manufacture of tin-plate (Section 5.6), the galvanization of iron and steel (Section 5.8), the alloying of iron with nickel (Section 5.9) and chromium (Section 5.10) to make stainless steels, and the processes used in the automobile plants to cover steel surfaces.

Iron is a constituent of some large biologically important molecules. One of these is hemoglobin which, as a constituent of blood, is able to carry dioxygen around the body. The iron is in its ferrous or II oxidation state, the atom having lost two electrons, and is centrally placed in a **complex** called heme. The bonding in complexes is described more fully in Chapter 11. A complex is formed between a metal ion and one or more molecules or ions which are called **ligands.** The metal ion - ligand(s) combination is called a complex. The metal ion and a ligand form a chemical bond by sharing two electrons. A covalent bond (Section 2.9) consists of two electrons being shared between two combining atoms, each atom contributing one electron. The bond in a complex differs from a covalent bond in that one partner; the ligand, supplies both electrons. This is called **co-ordinate** bonding, commonly represented by the symbolism:

$$M \leftarrow L$$

where M is the metal ion, L is the ligand, and the arrow indicates the donation of two electrons from the ligand to the metal ion. Heme is a complex of iron(II) with a ligand called porphyrin. Nitrogen atoms in compounds use three of their five valency electrons to form covalent bonds with three neighbouring atoms. Each nitrogen atom has a pair of electrons which may be used in the formation of a co-ordinate bond to a suitable acceptor such as a metal ion. In the heme complex, the iron(II) is bonded to four nitrogen atoms contained by the surrounding porphyrin ligand so that the five atoms constitute a square planar arrangement:

```
N       N
  ↘   ↙
   Fe
  ↗   ↖
N       N
```

In hemoglobin, a large protein molecule called globin acts as a fifth ligand and uses a nitrogen atom to donate two more electrons to the iron(II) of heme. This enables the globin molecule to surround the heme complex so that it permits only a restricted access to the iron(II). A sixth linkage to the iron(II) may be achieved with some difficulty by a water molecule as is the case with hemoglobin, or a dioxygen molecule. The replacement of water in hemoglobin (Hb) by dioxygen produces oxo-hemoglobin, HbO_2. The easily reversible interaction of hemoglobin with oxygen may be represented simply, omitting the water molecule, by the equation:

$$Hb + O_2 \rightleftharpoons HbO_2$$

The sign $\rightleftharpoons$ is used in chemical equations to signify that reactants and products are in a state of dynamic equilibrium, there being a facile interchange between the components of the system. It is important in the action of hemoglobin that the reversibility of the dioxygen addition reaction should be sufficiently rapid. There are actually four hemoglobin complexes in the complete hemoglobin structure so that it can carry up to four oxygen molecules. The local geometry of the globin protein allows the oxygen molecule to approach the iron(II) end-on and at an angle such that the Fe-O-O angle is about 130°. This effectively prevents the oxygen from causing the iron(II) to be permanently oxidized to iron(III), which is produced when iron(II) loses an electron. If the oxygen molecule had unrestricted access to the iron(II) centre it would cause permanent and irreversible oxidation to iron(III). The effectiveness of hemoglobin depends upon the bonding between the oxygen and the iron(II) being quite weak so that the oxygen may be transferred to another iron-heme-globin complex molecule called myoglobin (Mb) at the site of need in muscles:

$$HbO_2 + Mb \rightarrow Hb + MbO_2$$

Myoglobin proteins are stationary and act as an oxygen store. When the body needs energy the oxygen molecules are used in oxidation processes at the site of need.

The carbon monoxide molecule is similar in size to that of oxygen but has a much greater affinity for iron(II) than the latter by a factor of two hundred. The bond between carbon monoxide and the iron(II) of hemoglobin is strong and permanent. This is a major cause of the toxicity of the gas. The reactions of carbon monoxide with hemoglobin and oxo-hemoglobin are expressed by the equations:

$$Hb + CO \rightarrow HbCO$$

$$HbO_2 + CO \rightarrow HbCO + O_2$$

About six grams of iron are contained by an average person's body, two thirds of that, 4 grams, being in the blood. If all the iron in the hemoglobin were to react with carbon monoxide it would require about 2 litres of the gas at atmospheric pressure. After breathing in such a volume of carbon monoxide an average person would be dead, having no capacity to carry oxygen molecules to appropriate sites to provide the energy for life to be maintained. Carbon monoxide interferes with other iron-containing molecules which govern muscle activity, its reaction with hemoglobin not being the sole cause of death to people who inhale the gas. About 3 mL of petrol, if burned in a limited supply of air so that the main product is CO, would be sufficient to supply a lethal dose of the gas. In practice most of the petrol burned in car engines is oxidized to carbon dioxide and water but exhaust gases do contain between 1-12% of carbon monoxide as is evident from their use in suicides. Cigarette smoke contains about 4% of carbon monoxide, inhalers of this material subjecting themselves unnecessarily to chronic doses of the poisonous substance. The restricted access of molecules such as dioxygen and carbon monoxide to the iron centre in hemoglobin is important for the reversibility of the reaction of hemoglobin with dioxygen. If the access to the iron centre were to be unrestricted, the affinity for carbon monoxide would be some ten times greater than is actually observed. This would make even walking along the high street a very dangerous operation.

In addition to its contribution to hemoglobin and myoglobin, iron is a vital constituent of enzymes which are responsible for energy usage. Iron deficiency exhibits itself in the form of anemia, tiredness and listlessness. The recommended daily intake of iron is 10-15 milligrams, females needing more than males because of physiological differences between them. The main food sources of iron are nuts,

Jerusalem artichokes, peas, beans, lentils, potatoes (the iron is mainly in the skin), fruits and green vegetables. Cereals are usually fortified with iron and other minerals. Although it is an essential element, iron is toxic if taken at the level of 20 milligrams per kilogram of body weight. A dose of 1.6 grams of contained iron would be toxic to a person weighing 80 kg, such a dose being typically contained in 24 iron tablets.

5.3 COPPER

The abundance of copper in the Earth's crust is only 68 parts per million, i.e. 68 grams per tonne. Copper is the sixteenth most abundant metal. Its main ore minerals are chalcopyrite $CuFeS_2$, covellite CuS, chalcocite Cu_2S, bornite Cu_5FeS_4, and enargite Cu_3AsS_4. It is also found as the native metal and as the basic carbonate mineral called malachite, $CuCO_3.Cu(OH)_2$.

In 1990 the world production of mined copper ores was equivalent to 9.026 million tonnes of the metal with a value of $25 billion. Refineries produced 10.74 million tonnes of pure copper, indicating the use of some re-cycled scrap metal and reserves of ore. The distributions of the world production of metal content of ore and of refined copper metal are shown in Fig. 5.2. There are no other countries having a greater share than 3.5% of the copper ore production, but the 'Others' section of the percentage production of refined copper contains significant contributions from Japan (9.4%), Germany (4.4%) and Belgium (3.1%), countries which have no ore production, but have refineries using imported materials. The U.K. has no remaining workable deposits of copper ore, but produced 1.1% of the world's refined metal in 1990. The total refined copper metal produced in 1990 had a value of $25 billion.

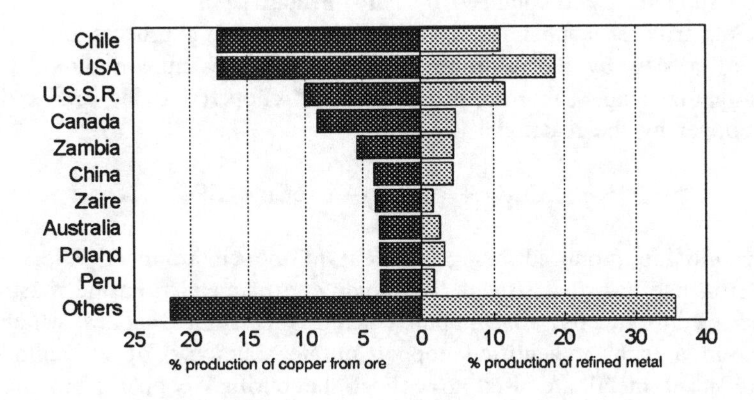

Fig. 5.2 The distributions of the World production of copper metal from ore and of refined copper metal in 1990

The 'Others' sections of Fig. 5.2 indicate that the production and refining of copper is carried out in a large number of countries, each contributing on a minor scale to the totals.

The Chuquicamata mine in Chile has the greatest production of copper in the world (680,000 tonnes in 1990) from what is regarded as the richest and largest known deposit in the world. In 1990 the mine measured 3.9 km long, 2 km wide and 610 m deep. Possibly the largest hole in the world is at the Kennecott (RTZ owned) mine at Bingham Canyon, near Salt Lake City, and which is shown in the cover photograph. The copper deposit consisted originally as a mountain some 275 metres above the valley floor. Sanford and Thomas Bingham, the brothers who gave their name to the town in the valley, were sent to the area by Brigham Young, the founder of the Mormon religion, in 1848. Although Brigham Young knew that some mineral wealth existed in the Canyon, he insisted that the area should be developed as a forest and farming community. He did not wish to provoke an influx of gentile prospectors who would not be interested in his aims of '*raising wheat, oats, barley, corn, and vegetables and fruits in abundance so that there may be plenty in the land*'. After the Civil War, a gentile from an Irish immigrant family, Patrick Connor (the family had dropped the O'), eventually established Bingham Canyon as a mining district in 1863. In 1991 the output from the Bingham Canyon mine amounted to 236,500 tonnes of copper, which was 15% of the US production, 14.24 tonnes of gold (4.7%) and 7000 tonnes of molybdenum (12%). The town of Bingham no longer exists, having been destroyed to gain access to the valuable deposits on which it was sited and the original mountain has been transformed into a hole around 800 metres deep and 4 kilometres in diameter. More than 5000 million tonnes of material have been removed from the open pit, the current rate of removal being two hundred thousand (200,000) tonnes every day. Some people may regret the conversion of a natural mountain into a large hole, but without copper there would be no production or transmission of electricity and the great organ in the Mormon Tabernacle in Salt Lake City would have to be supplied by air pumped by hard manual labour.

The majority of copper ore is treated by a roasting process in which iron is removed as a slag by the addition of silica, which is silicon dioxide SiO_2, and found as quartz sand. The resulting mixture of copper(I) oxide and sulfide yields metallic copper by the reaction:

$$2Cu_2O + Cu_2S \longrightarrow 6Cu + SO_2$$

the sulfur dioxide produced being a threat to the environment unless chemically removed from the gaseous effluent. The crude copper metal from the roasting process is refined electrolytically. The impure metal is cast into anodes which are then suspended in a bath of acidified copper sulfate solution. Copper cathodes formed from pure sheet metal are used to collect the purified copper from the anode as electric current is passed through the electrolytic cell. Electrons are delivered to the cathode which attracts copper(II) ions from the copper sulfate solution and discharges them as neutral copper atoms on the cathode. Electrons are removed from the anode which causes the release into solution of an equal number of copper(II) ions. The refined copper metal from the cathodes is then ready for its various uses.

The sludge deposited in the electrolytic cells is a rich source of gold and platinum group metals, i.e. ruthenium, rhodium, palladium, osmium, iridium and platinum. The sale of the sludge covers the cost of the electricity used in the refinement process.

An ever growing percentage of copper is produced either from the tailings and waste dumps of mines or by *in situ* leaching. In these cases the metal is leached from the ore in the dump or in the deposit itself by the use of acidic mine water. Acidic mine drainage is water which percolates through mine workings and becomes acidified because of bacterial action upon sulfide ores. Rock-eating bacteria called chemolithoautotrophs are now well characterized and derive their energy by oxidizing sulfides and iron(II) minerals. Their sole source of carbon is carbon dioxide and they use oxygen for the oxidation of their chosen 'foods'. Their action has only recently been understood, but the products of their action have been known about ever since mines were invented. The Romans made use of the process without understanding it and there is a graphic description of it in a fictionalized version of the history of industry in Anglesey in the early nineteenth century by Alexander Cordell in his book, *The Land of my Fathers.*

'Those were the early weeks I spent working at the Precipitation pits, and received seven shillings a week, the wages of a man. Marvellous are the works of humans to take red water from a mountain and turn it into copper.

See them digging the new shaft on Parys - red-core Parys that is riddled with copper - working as did the Romans. Down, down goes the shaft into the bowels of the mountain: fill the shaft with water and leave it for nine months: then pump it out into Precipitation.

In come the ships carrying scrap iron for the stew.

Aye, this is the stew of copper - old bedsteads, old engines, wheels and nails - scrap of all description from the iron industry of the North comes surging into Amlwch on the ships. The carters unload it and carry it to the pits: and in it goes, iron scrap into the sulphate stew, the blood of Parys. And we, the stirrers, worked the brew with wooden poles - wooden ones, since metal ones would melt in the acid - even copper nails in our boots. Stir, stir, stir - nine months it took to melt the iron into a liquid. And the sludge that dropped to the bottom of the pits was copper. Break this up cold and cart it like biscuits down to the maw of the Smelter for refining.

This was the process; a job for men, not boys.'

The chemistry of the process may be represented by the equations:

$$4FeS_2 + 15O_2 + 2H_2O \longrightarrow 2Fe_2(SO_4)_3 + 2H_2SO_4$$

$$CuS + 2O_2 \longrightarrow CuSO_4$$

The iron pyrites, FeS_2, is oxidized to iron(III) sulfate and sulfuric acid, and the copper sulfide is oxidized to copper(II) sulfate. The iron(III) and copper(II) sulfates are soluble in the very acidic solution. The use of scrap iron to produce copper metal may be written as the equation:

$$Cu^{II} + Fe^{0} \longrightarrow Cu^{0} + Fe^{II}$$

The copper metal is produced as a sludge which is separated from the rest of the material and sent to the refinery. The iron metal displaces the copper ions, dissolving in the form of iron(II) ions and producing the copper as the metal, Cu^0.

The same process is being used currently in the USA to produce about 15% of the total copper production. Bacterial oxidation, as a metal extraction process, is very much more environmentally acceptable than the alternative roasting method. No atmospheric pollution is caused and the aqueous waste products are safely stored as solid precipitates; hydrated iron(III) oxide and iron(III) arsenate, if any arsenic is present.

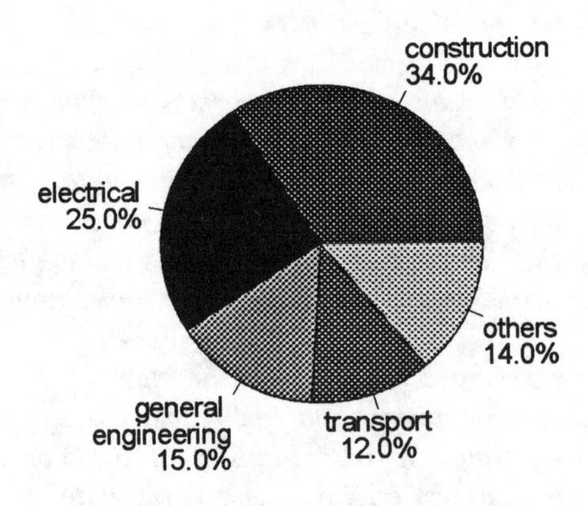

Fig. 5.3 The distribution of the uses of copper

The distribution of the uses of copper is shown in Fig. 5.3. The metal is used for the transmission of electricity because of its high electrical conductivity and its ductility, the ease of its formation into wire. It is incorporated into various alloys which include brass (20-50% zinc), bronze (1-10% tin, 0.35% phosphorus in phosphor bronze) and 'nickel silver' (55-65% copper, 10-18% nickel and 17-27% zinc). The main uses of copper in the home are in the pipework of plumbing and central heating, the hot-water cistern, the electrical wiring of the house and the electrical appliances used therein. Car radiators and wiring are made from the metal. Low denomination coins are mainly copper, 'silver' coins being made from alloys of copper and nickel. From 1993 onwards, British 'copper' one and two penny coins have been manufactured from steel with a copper coating, a cheaper combination than if the coins were composed entirely of copper. The new coins could give the false impression that copper is magnetic! In the form of compounds, copper is used for insecticidal and fungicidal purposes in agriculture. What would our quality of life be without copper?

Copper is an essential element for human life: the body requires an intake of about 2-3 milligrams per day to maintain its normal content of 60 mg. The element controls the storage and release of iron in hemoglobin, is important in enzymes which are used in food metabolism, and is a constituent of the sheath which covers nerve tissue. The element is very widespread in foods and no supplements are recommended, no cases of copper deficiency having been reported. An ingestion of too much copper can cause illness or even death in a human being, 1-2 grams being sufficient to be lethal.

Clean copper metal has a characteristic pink 'copper' colour, but when exposed to the moist atmosphere for some time becomes coated with a layer of oxide (CuO) which is black. This eventually reacts with atmospheric carbon dioxide and water to give the light green basic copper carbonate with the same formula and colour as the mineral, malachite. Copper sheeting, used as a roofing material, soon acquires a coating of the green compound. Copper exists in aqueous solution as Cu^{II}, the Cu^{2+} ion, which has a characteristic blue colour. This may be produced by leaving a small piece of copper tubing in household ammonia solution overnight. In the alkaline conditions dioxygen oxidizes the copper to its II state which then forms the complex, $[Cu(NH_3)_4]^{2+}$, the four ammonia molecules acting as ligands to the central Cu^{2+} ion. The deep blue complex is square planar, as is the arrangement around the iron(II) in heme.

5.4 SILVER

Silver is a relatively rare element, present in the Earth's crust to the extent of only 0.08 grams per tonne. It occurs as the native element, often in association with gold, and as the sulfide mineral argentite, Ag_2S, and the chloride mineral called horn silver, AgCl. The sulfide is associated with deposits of the sulfide ores of lead, zinc and copper, the metal following the paths of extraction of these base metals and being a by-product of their production.

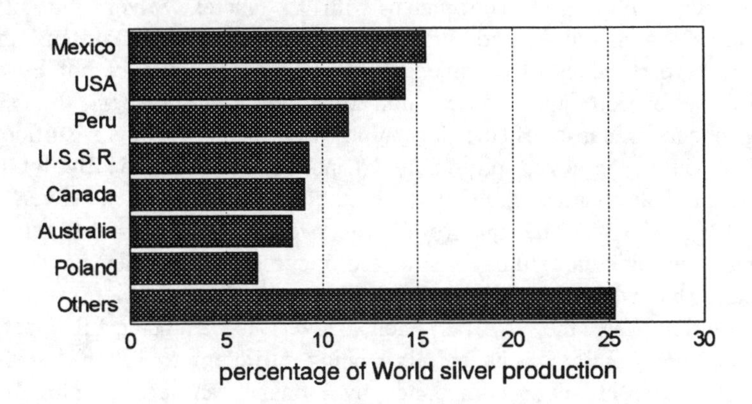

Fig. 5.4 The distribution of the World production of silver in 1990

Total world production of silver in 1990 was 15108 tonnes of metal worth $2 billion. The distribution of the world production in 1990 of silver is shown in Fig. 5.4. The 'Others' section of the diagram indicates that many other countries contribute relatively small amounts to the total production.

Silver is produced from its sulfide ore by reduction with carbon, but the majority of its production is from the processes in which other metals are the main product. It follows the same route as does gold in the cyanidation process used for that metal (Section 5.5). It is produced from the anode slimes in the electrolytic refining of copper and zinc and is extracted from lead by treating the molten metal with zinc. In the last process the silver is extracted into the zinc which forms a separate layer above the lead, the two metals being immiscible in their liquid states.

Silver is a relatively valuable metal and, though it is little used for coinage now, it is stored by governments and by private investors, recent totals being 8650 and 40000 tonnes respectively, as a commodity. Its main use is in photography as the halides, silver bromide (AgBr) and silver chloride (AgCl), a halide being a compound of a metal with any Group 17 element (they are called halogens; from the Greek meaning salt-formers). The two compounds are sensitive to light, the action of which activates the irradiated material, which is then converted into metallic silver at the development stage. The main constituent of the developer solution is hydroquinone, an organic compound with powerful reducing properties. The activated silver halides are reduced to the metal and the excess of the halides, i.e. those in the parts of the film which have not been exposed to light, is dissolved by sodium thiosulfate solution (called hypo) at the fixing stage. Of the silver not used for investment purposes, around 50% is used for photography. Electrical components account for another 25% of the uses of the metal. Ten percent is used in the production of silverware and in electroplating, 5% being used in the form of brazing alloys and solders.

Silver is important in the home for photographic purposes and for the silvering of mirrors, jewellery and decorative articles, and items of cutlery which are either solid silver or which are electroplated with the metal. Silver and silver plated objects tarnish with time and have to be polished to maintain their shiny appearance. In particular, silver objects and cooked eggs should not be allowed to meet. The sulfur-rich egg causes immediate discoloration of the silver. The tarnishing occurs because of the formation of a layer of silver sulfide which is black. This can be removed physically by polishing or chemically by the use of dilute household ammonia which dissolves the silver sulfide as a silver(I)/ammonia complex; $[Ag(NH_3)_2]^+$. The quality of life would be poorer without silver; no photography, no cinema, although still and moving images can now be produced electronically, but which need copper and silicon for their production.

In solution, as the $Ag^+(aq)$ ion, silver is a powerful bactericide, a concentration of 10 micrograms per litre being sufficient to kill bacteria. Although the ingestion of solid silver compounds by humans can lead to chronic skin and pulmonary diseases, silver ions in solution, taken by mouth, are converted to insoluble silver chloride in the stomach and do not lead to major toxicity.

5.5 GOLD

Gold is found mainly as the native element, but is of low abundance, accounting for only 0.004 grams per tonne of the Earth's crust. Gold ores contain around 4 grams per tonne of the metal, 4 grams of gold being contained in a typical wedding ring. Its 'nobility', its resistance to oxidation, coupled with its attractive appearance and scarcity, have combined to make it the most desired element. The world production in 1990 was 2048.6 tonnes of the metal worth $24 billion. The distribution of world production in 1990 is shown in Fig. 5.5. Many other countries contribute relatively small amounts to the total production.

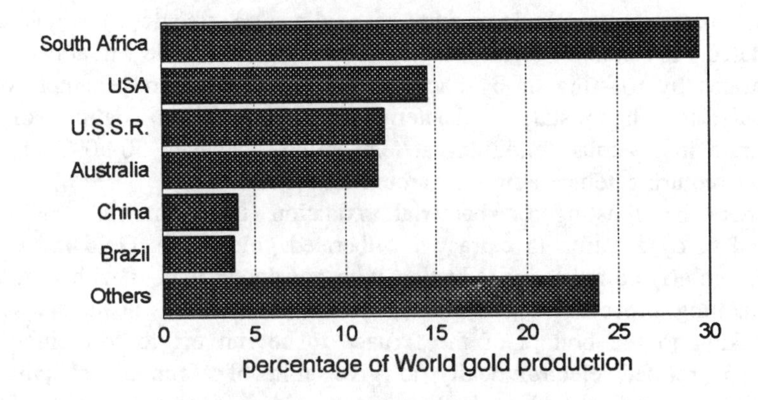

Fig. 5.5 The distribution of World production of gold in 1990

Although very large nuggets of the native metal have been found in the Earth, the majority of current gold production originates from microscopic grains of the metal. Some of the grains are encapsulated by sulfide minerals such as iron pyrites, arsenopyrite FeAsS, and chalcopyrite $CuFeS_2$. The normal method of extraction of gold is by cyanidation. The crushed and ground ore is treated with a solution of sodium cyanide which is made to have a pH value of 10 by the addition of sodium hydroxide. The presence of the hydroxide ion prevents the production of volatile hydrogen cyanide which is intensely poisonous. The cyanide ions react with gold metal in the presence of atmospheric dioxygen to form a soluble gold cyanide complex. The gold metal is oxidized to its (I) state by dioxygen and is stabilized by forming a gold(I)-cyanide complex ion:

$$4Au + O_2 + 8CN^- + 2H_2O \longrightarrow 4[Au(CN)_2]^- + 4OH^-$$

Square brackets are used to enclose the formulae of complex ions in which metal ions and ligands are linked via co-ordinate bonds. In the case above, the complex ion, $[Au(CN)_2]^-$, consists of a central gold(I) ion linked to two cyanide ions, the overall ionic charge being -1. The solubility of gold in cyanide solutions was first reported by the Swedish chemist Scheele in 1783, but has been used in the gold industry only from around 1887 when three Scots, McArthur and the Forrest brothers, patented the process.

Cyanide ions cannot penetrate the coating of sulfide minerals that some gold particles possess. Such material is concentrated by froth flotation and may be roasted or be subjected to bacterial oxidation to remove the sulfide mineral. Roasting is the conventional process but produces gaseous sulfur dioxide which is an atmospheric pollutant and solid finely divided particles of arsenic(III) oxide, if arsenic is present in the sulfide minerals, which requires containment to protect the environment. The advantages of bacterial oxidation are that no atmospheric pollutants are produced, the sulfur being oxidized to sulfuric acid which can be neutralized with slaked lime (calcium hydroxide, $Ca(OH)_2$) and removed as insoluble calcium sulfate. Any arsenic present is oxidized to arsenic(V) which can then be precipitated in stable form as iron(III) arsenate. The treatment of sulfide concentrates by roasting or by bacterial oxidation is a good example of the effect of catalysis. The catalytic bacteria facilitate the oxidation of a sulfide concentrate in aqueous conditions at temperatures between 30-50°C, the roasting of the solid requiring temperatures of around 600-800°C.

After the roasting or bacterial oxidation treatment, the solid residue is subjected to cyanidation to extract the liberated gold metal. Gold metal is produced from the gold(I) complex by reduction with powdered zinc or with iron wool. In the final smelting process the base metals are converted to silicate slags, the molten gold sinking to the bottom of the furnace to be run off to form ingots. The gold ingots are refined electrolytically to give gold of 'four nines' purity (99.99%) which is the product sold as bullion. Pure gold is considered to be 24 carat (the carat is the name of the unit of weight used by diamond merchants, one carat being 0.2 grams), gold used for jewellery being produced as alloys which are 22, 18 or 9 carat, with those numbers of twenty-fourths of the precious metal. The metals alloyed with gold are usually copper or silver which have an influence upon the colour of the gold alloy. The addition of copper increases the 'yellowness' of the gold, the addition of silver producing 'white gold'.

Around 12% of gold production in 1990 was bought for investment purposes by governments and private companies and individuals. The remainder of the gold available, including that bought from investors and scrap, was used principally by the jewellery industries (73%) with electronic industries (14%), dentistry (5%) and coin and medal production (5%) accounting for considerable quantities. The individual figures for uses in electronics industries in Japan (68 tonnes or 33% of total Japanese usage) and the USA (38 tonnes or 19% of total USA usage) are indicative of where the world's electronic goods are mainly manufactured. The EEC countries used 24 tonnes of gold in electronics manufacture which was only 4% of their total usage. It is noteworthy that about 88 thousand tonnes of gold, worth $930 billion at today's price, are stored away in vaults of banks. The sophisticated processes that are used to liberate and extract the gold from the Earth result in great stacks of ingots of the pure metal which are stored again deep in the Earth! It has been estimated that around one hundred thousand tonnes of metallic gold has been produced. This amount of the metal would occupy a cube with sides of just over seventeen metres. The stored gold represents a hedge against the time when people and governments lose confidence in paper money.

In the home, gold is not very important except in computer keyboard contacts and in symbolic and decorative jewellery. Two of the author's back teeth depend upon gold alloy caps for their continuing stability and effectiveness.

Because of the nobility of metallic gold, i.e. its resistance to oxidation, a combination of a powerful oxidant and a complexing agent is required to render it soluble. The metal will react with fluorine and chlorine to give gold(III) compounds, in the absence of water. Apart from the oxygen gas/cyanide ion combination which produces the gold(I)-cyanide complex the most common solution which is used to dissolve gold is *aqua regia*, a 3:1 by volume mixture of concentrated hydrochloric (HCl) and nitric (HNO_3) acids. The concentrated nitric acid provides the oxidizing power, the gold(III) produced being stabilized as a gold(III) chloride ion complex, $[AuCl_4]^-$.

5.6 TIN

The abundance of tin in the Earth's crust is 2.1 grams per tonne. The element occurs mainly in the form of the oxide mineral, cassiterite SnO_2, and to a very small extent as the mixed copper/iron/tin sulfide, stannite Cu_2FeSnS_4. The metal is produced from its oxide ore by using carbon, in the form of coke, as a reducing agent. Stannite is a minor constituent of the ore produced by some copper mines and is separated during the extraction process of the major metal. Tin cannot be extracted from stannite economically currently, any extracted mineral being stockpiled. It is possible that bacterial oxidation of the mineral to solubilize the metal content will become economic in the near future.

The production of tin metal in 1990 amounted to 210.7 thousand tonnes worth \$1.2 billion. Recycled metal production in 1990 was in the region of 18 thousand tonnes, this being a low estimate because of the lack of data from some countries.

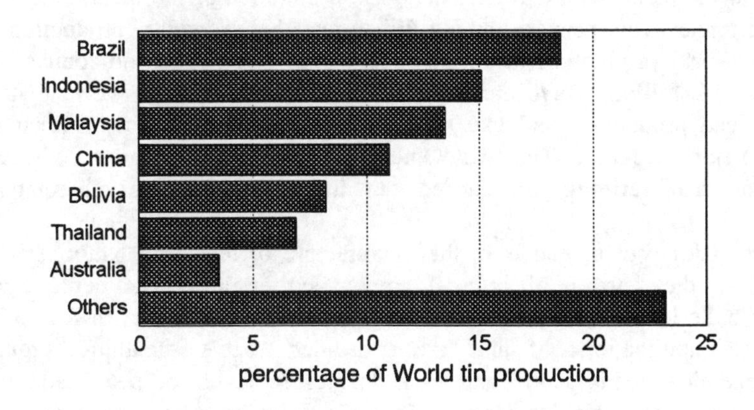

Fig. 5.6 The distribution of World production of tin in 1990

The distribution of world production of the metal, excluding any recycled material, is shown in Fig. 5.6. Tin production is widespread as indicated by the large contribution from 'Other' countries, which each supply relatively small amounts to the total production.

The main uses of tin are in the production of tinplate (31%), solders (these are alloys of tin with lead and some minor quantities of other metals) (26%), alloys such as bronze (containing between 5 and 10% tin as an additive to copper) (16%) and the manufacture of various chemicals including fungicides, biocides and fire retardants (28%). Molten tin is used in the production of float-glass, the level surface of the liquid metal allowing the manufacture of glass with a flat surface. Tinplate is common in the home, but is being replaced rapidly by aluminium as the containers of foodstuffs. The cans of beloved and nutritive foods such as baked beans and mushy peas are lacquered internally to avoid any dissolution of the metal. The more acidic foodstuffs, e.g. rhubarb, would cause appreciable reaction if the the cans were not lacquered. The soldering of all domestic appliances and electronic goods is dependent upon tin. The use of tin in tinplate is to prevent the oxidation of the iron which is covered with a very thin layer of the metal (up to 25 μm thick), tin itself being resistant to oxidation. Tin(II) fluoride was used in some toothpastes to provide the fluoride ions which were incorporated in the teeth for strengthening and the prevention of decay. This has now been replaced by sodium fluorophosphate which allows a more efficient transfer of fluoride ion to the teeth. Pewter, which is an alloy of 90-95% tin with minor proportions of antimony and copper, is used for ornaments, trophies and drinking vessels such as tankards. There seems to be no substitute for tin in solders so it is an element which is highly essential for modern life.

5.7 LEAD

The abundance of lead in the Earth's crust is 13 grams per tonne. The main ore is the sulfide mineral, galena, PbS. The sulfide is roasted to give the oxide which is reduced to the metal with carbon. The world production of the metal in 1990 was 3316 thousand tonnes from primary mining. When recycled secondary production is included the total world production of refined metal was 5674 thousand tonnes. This had a value of $3.2 billion. There are some differences between the world distributions of primary lead production and that of refined lead. These data are shown in Fig. 5.7 for the major producers. The two 'Others' sections of the diagram indicate that lead production and refining are carried out in many countries on relatively minor scales.

The major use of lead is in the manufacture of lead-acid batteries which are so essential to the starting of internal combustion engines. The percentages of lead used for batteries in the car manufacturing countries are shown in Fig. 5.8.

Minor applications of lead are in solders, cable sheathing, piping, roofing sheets, chemical manufacture and to a decreasing level for petrol additives. In the home lead is found in the solders of domestic appliances and electronic goods and is essential to their actions. Lead is toxic and may be ingested from such sources as artists materials, i.e. glazes and oil paints, paint chips from old objects, most modern paints being lead-free, and solders. There are products which are marketed with the purpose of turning grey hair to a black colour. They consist of dilute solutions of lead acetate. Such solutions contain the aqueous lead(II) ion, $Pb^{2+}(aq)$, which reacts with some of the sulfur content of the hair (the amino-acids cysteine and methionine containing sulfur) to produce lead sulfide, PbS, which is a black solid. It is this potentially poisonous substance which gives the treated hair its dark hue. If any of the lead acetate solution or the lead sulfide were to become ingested by the body it would cause some toxicity. Chronic lead poisoning leads to permanent mental retardation and anemia, larger doses leading to coma and death.

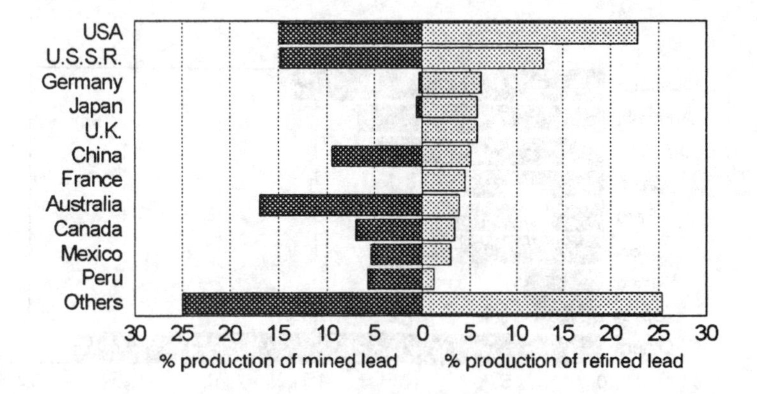

Fig. 5.7 The World distributions of primary lead production and that of refined lead in 1990

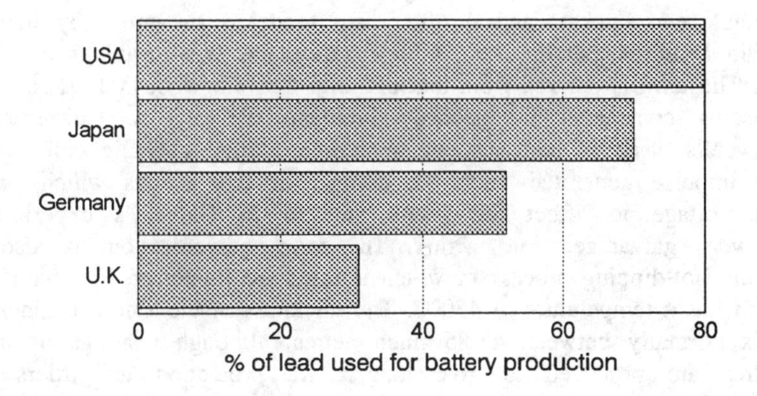

Fig. 5.8 The percentages of lead used for batteries in the major car manufacturing countries

5.8 ZINC

The abundance of zinc in the Earth's crust is 76 grams per tonne, its main ore minerals being the sulfide, zinc blende ZnS, and the carbonate, calamine $ZnCO_3$. The sulfide is roasted to give the oxide, ZnO. The action of heat upon the carbonate mineral gives the oxide. The oxide is then reduced to the metal with carbon. The total world production of zinc in 1990 was 7.3 million tonnes worth $10.8 billion. The distribution of world production of the metal in 1990 is shown in Fig. 5.9. As indicated by the 'Others' section of the diagram, zinc production is widespread.

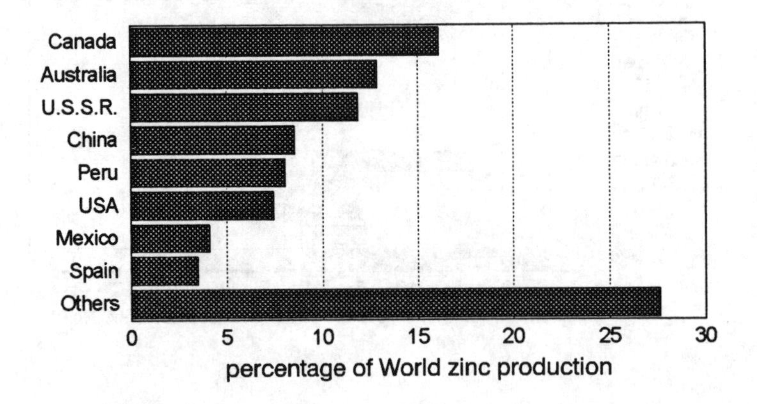

Fig. 5.9 The distribution of World production of zinc in 1990

The main use of zinc is in the galvanization process in which iron and mild steel articles, e.g. buckets and dustbins, are rendered rust-proof by being coated with a thin layer of metallic zinc. Galvanization has two meanings in the English language. The term is derived from that of Luigi Galvani who, in 1791, observed that frog muscles contracted if touched simultaneously by two different metals. Electrical cells, used in batteries, are sometimes called galvanic cells. It was the electrical impulse generated from the contact of two metals which produced a sufficient voltage to affect the frog's muscles in Galvani's experiments. The muscles were galvanized into action. The term, galvanization, is also used to describe the hot-dipping process in which articles are immersed in a bath of molten zinc which has a temperature of 450°C. The thickness of the zinc and zinc/iron alloy coating is normally between 45-85 micrometres, although coatings of up to 250 micrometres are produced to give longer life protection to girders used for constructional purposes. Between the outer layer of zinc and the steel article there is an intermediate layer which is a zinc/iron alloy. The treatment produces interesting patterns of crystalline zinc on the surface of galvanized objects. The zinc layers protect the steel from oxidation even if the steel becomes exposed to the atmosphere, zinc being oxidized more readily than iron. Other uses of zinc are in the production of die-casting alloys and brass. Metallic zinc forms the negative pole of zinc-carbon and alkaline manganese battery cells. In the home it is used in galvanized articles (those now being largely replaced by polypropylene products), batteries and in common pharmaceutical products such as zinc ointment, zinc plasters and calamine lotion. The pharmaceutical products make use of the bactericidal properties of zinc(II).

Zinc is an essential element for the human body, which contains around 2 grams of the element. Around 70% of the zinc is contained by bone, the remainder contributing to many enzymes, e.g. insulin, which is a zinc complex. Zinc is essential for proper growth, the healing of cuts and wounds, the digestion and metabolism of food, and is important in the mechanisms of the sensations of smell and taste. Whole grains, peas, beans and nuts are good sources of zinc, particularly for vegetarians who have a slight risk of zinc deficiency. Persons enjoying a normal diet, whether vegetarians or otherwise, very rarely suffer from this as the element is fairly widely available in foods. The recommended daily intake is 15 milligrams but it not normal to supplement the diet. A dose of 500 mg can cause vomiting.

5.9 NICKEL

The abundance of nickel in the Earth's crust is 99 grams per tonne, its main ore mineral being pentlandite, $(Fe,Ni)_9S_8$. The formula indicates that the iron and nickel content is variable but should amount to a total of nine metal atoms for every eight atoms of sulfur. Nickel is produced in various parts of the world, the world production of the metal in 1990 being 874 thousand tonnes worth $7 billion. The distribution of world production in 1990 is shown in Fig. 5.10. The 'Others' section of the diagram indicates that many countries contribute relatively small amounts to the total production.

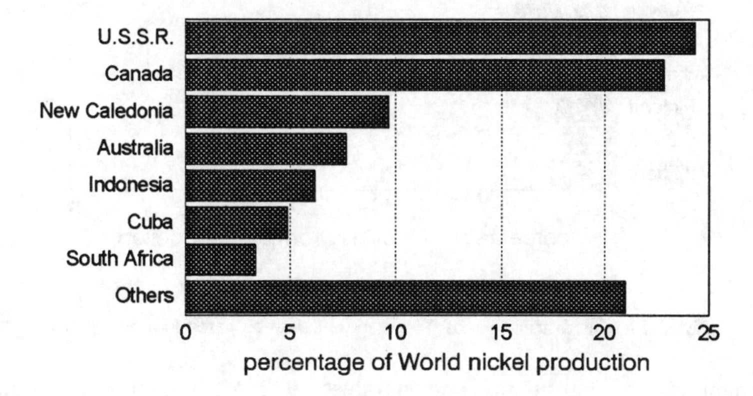

Fig. 5.10 The distribution of World production of nickel in 1990

The main use of nickel is in the production of stainless steel, which contains around 8% of the metal, and other alloy steels. About 15% is used for electroplating purposes. It is used in the production of specialized magnets and electrical battery cells, particularly those which are rechargeable. It forms 8% of stainless steel cutlery and the kitchen sink.

5.10 CHROMIUM

Chromium has an abundance in the Earth's crust of 122 grams per tonne. The only commercially useful source is the mineral, chromite $FeCr_2O_4$, a mixed iron(II)/chromium(III) oxide. Very little pure metallic chromium is produced since it is usually used in stainless steel production where the iron content of the chromite is also used. The reduction of chromite to ferrochrome is achieved by the use of carbon (coke) in an electric arc furnace. This iron/chromium alloy is added to steel, together with the required amount of nickel, to produce stainless steel. Stainless steel usually has about 18% of chromium metal content.

The total world production of chromium in 1990 was equivalent to 3.8 million tonnes of the metal. The value of this in terms of chromite mineral was $700 million. The distribution of world production in 1990 is shown in Fig. 5.11, but data from Bulgaria, North Korea and Thailand are not available, although those countries do have some production. Unlike all the metals considered in the previous sections, there is only a small contribution from countries other than those mentioned in the Figure. The occurrence of economically available chromium minerals is highly localized, a factor which gives the metal some strategic importance.

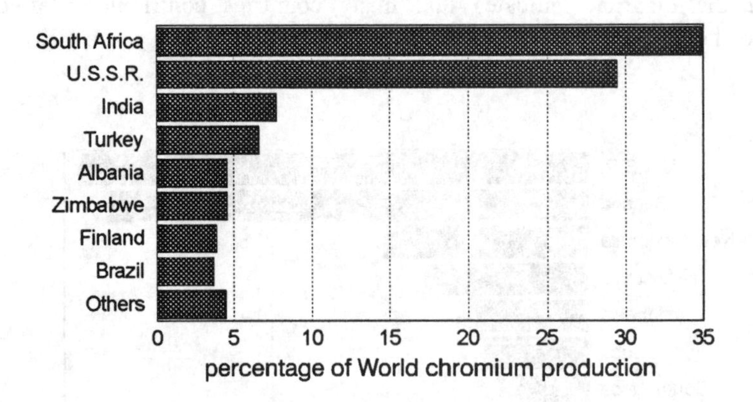

Fig. 5.11 The distribution of World production of chromium in 1990

The main use of chromium is in stainless steel production with electroplating being a relatively minor use. The 'chrome' used for trimming cars is steel with 17% chromium content. The effect of high chromium content of steels is cause them to be resistant to oxidation. The chromium atoms at the surface become oxidized to give a coherent and substantial layer of oxide which is not permeable by water. This is the difference from iron and non-stainless steels where the initially formed oxide layer is permeable and allows the rusting to proceed throughout the piece of metal.

It has been established that trace quantities of chromium are essential for the maintenance of human metabolism of sugar. The so-called glucose tolerance factor, a compound containing chromium, enables insulin to effect the metabolism of sugar in the blood stream. A chromium deficiency can produce the illness known as late-onset diabetes. It is common for new mothers to exhibit temporary diabetic symptoms as they may have a chromium deficiency as a result of passing too much of their body store of the element on to their foetal child during pregnancy. Chromium deficiency is also implicated in the development of hardening of the arteries, known as atherosclerosis. The element is considered to have a role in the synthesis of fatty acids and cholesterol.

Most foods, i.e. meat, vegetables and fruit, contain some chromium but the highest concentration is found in the fractions of cereals which are present in whole food but which are not present in more refined foods. Relatively very high concentrations of chromium are found in black pepper, the outer covering being important as it is with cereals, and spices in general, blackberries, and in wines of any colour. The chromium content of wines, and beers to a lesser extent, is thought to be connected with the yeast which causes the fermentation process and which grows in conjunction with the vines. Distilled alcoholic drinks, i.e. spirits, have no chromium content because the chromium remains in the residue of the distillation process. Brewer's yeast is an excellent source of the element. An even better method of acquiring sufficient chromium would be to indulge in a good chromium-rich lunch. This might be a chicken sandwich with lots of black pepper using whole meal bread, washed down with a glass of wine, followed by a bowl of stewed blackberries with another glass of wine. If the blackberries were stewed in a stainless steel pan so much the better, since chromium is leached out slightly in the stewing process. There cannot be a better preventative medicine than this!

5.11 ALUMINIUM

Aluminium is the most abundant metallic element, accounting for 83 kilograms per tonne (8.3%), of the Earth's crust. The main ore mineral is bauxite which is a mixed oxide/hydroxide of very variable composition. The two extremes have the formulae $AlO(OH)$ and $Al(OH)_3$, but real deposits lie somewhere in between. The metal is extracted by electrolyzing a molten mixture of the oxide (Al_2O_3) and cryolite (Na_3AlF_6). The oxide is produced by purifying bauxite ore, the cryolite being produced synthetically. The cryolite is important in that it allows the electrolysis to occur at a temperature of around 960°C, the melting point of aluminium oxide being 2072°C. The oxide is dissolved in molten cryolite and kept at a concentration of around 8% as aluminium metal is discharged at the cathode. Cheap electrical power is essential for economic production of aluminium, the metal usually being extracted wherever there are sources of hydroelectricity. The total world production of aluminium in 1990 was 18 million tonnes which was worth $24 billion. The distributions of world bauxite production and that of refined aluminium metal, in 1990, are shown in Fig. 5.12. The 'Others' sections of the diagram indicate that, although 20% of the total bauxite production is contributed by many other countries, the production of refined aluminium is even more widespread.

About 60% of aluminium metal produced is used in packaging, including cans, building, and transport, especially in aircraft. The remainder is used for electrical power lines, the manufacture of consumer durables and in machinery. In the home the metal is obvious in packaging of take-away foods, kitchen foil, kitchen utensils, e.g. pans and steamers, and in some window frames. There is some doubt about the use of aluminium pans because of the possible link of intake of the metal and Alzheimer's disease. There is a great amount of circumstantial evidence for the link and it is advisable not to use aluminium cooking vessels until the issue is resolved. In particular acid foods such as fruit, especially rhubarb, should not be cooked in aluminium pans.

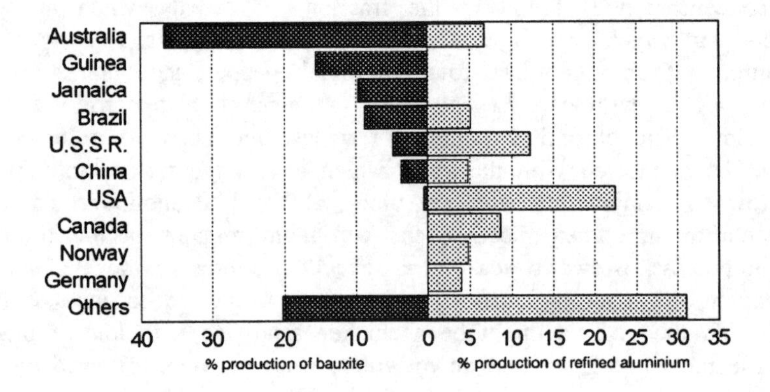

Fig. 5.12 The distributions of World bauxite production and that of refined aluminium metal (1990 data)

The chemistry of aluminium is that of the metal and its (III) combined state. The metal is, in theory, easily oxidized since it is quite electropositive. In practice, a thin coherent and resilient film of aluminium oxide protects the metal from any further oxidation. Only if this protective oxide layer is damaged does the metal oxidize further. In its crystalline form, the oxide is found as the mineral known as corundum which is colourless when pure, and is exceptionally hard. Only diamond, of the naturally occurring minerals, is harder. Corundum is very rarely found in a pure form, but the impure versions which do exist are prized as gems. Rubies are corundum crystals with a chromium(III) impurity, sapphires being corundum crystals tainted with titanium(III). The gems known as oriental topaz, oriental amethyst and oriental emerald are other versions of impure corundum, although topaz which is a basic aluminium fluorosilicate, amethyst which is quartz or silicon dioxide, and emerald which is beryllium aluminium silicate with a chromium impurity, are chemically unconnected with it. Emery, an industrial grade of corundum with an iron impurity, is used extensively in the manufacture of abrasives.

The inconvenience caused by the absence of aluminium foil and cans from everyday existence is obvious, but not crucial. That caused by there being no air travel would be considerable; without aluminium there would be no planes, except for Concorde which is mainly constructed from titanium.

5.12 METAL RESERVES

The reserves of the metals discussed in this chapter are well defined, but will vary with time as new deposits are found to be economically exploitable. The amounts of known exploitable reserves for each metal may be expressed in terms of the number of years of production, at present day levels, which will cause their depletion. These data are given for the metals discussed in this chapter in Table 5.2, together with an indication of the three countries possessing the most reserves, and the percentage of the reserves possessed by each of the three countries.

Table 5.1 - The lifetime of current reserves of some metals, the three main countries where the reserves are situated and the percentage of those reserves possessed by the countries

metal	lifetime /years	countries having most mineral reserves and % possessed
copper	36	Chile(27), USA(17), U.S.S.R.(12)
silver	19	U.S.S.R.(16), Mexico(13), Canada(13)
gold	22	S.Africa(46), U.S.S.R.(15), USA(11)
iron	119	U.S.S.R.(37), Australia(16), Canada(7)
tin	28	China(25), Brazil(20), Malaysia(19)
lead	20	Australia(20), USA(16), Canada(10)
zinc	21	Canada(15), USA(13), Australia(12)
nickel	55	Cuba(38), Canada(13), New Caledonia(10)
chromium	105	S.Africa(70), Zimbabwe(10), U.S.S.R.(10)
aluminium	220	Guinea(26), Australia(21), Brazil(13)

There is an important difference between the reserves of metallic elements and those of the non-renewable fossil fuels. The extent of fossil fuel reserves will, as will those of the metals, increase as more deposits become economically exploitable. Eventually, in the distant future, i.e. 50-200 years, the fossil fuels will run out if they are used at current rates. This does not apply to metals which, although they go through the processes of extraction and use, are forever available to be re-used by recycling or even by re-extraction from waste materials. More attention should be paid to the re-cycling of metals and to the conversion of metal-containing waste into useful forms.

6

Fire and the fossil fuels

Formation of natural gas, petroleum and coal. Energy equivalents. Distribution of world production. Chemical nature. Processing. End uses. Reserves and resources. Economic groupings of countries of the World. Economics of domestic use of fossil fuels compared to those of electricity usage. Alternative sources of energy.

6.1 THE FORMATION OF FOSSIL FUELS

Our main sources of fire are the fossil fuels which have an organic origin. These are natural gas and petroleum which were formed from the decay of marine organisms, and coal which was produced by the decay of trees. The trees were generated photosynthetically from carbon dioxide and water, the living organisms themselves having lives dependent upon photosynthesized food. The energy content of the fuels originated in the Sun and was transmitted over the 147 million kilometres of space to the Earth, the same source of energy as all life depends upon today. Since natural gas, petroleum and coal are non-renewable resources they should be used prudently. Whether they are being used prudently is a matter of reasonably urgent debate. The information contained in this chapter is relevant to that debate.

In very simplified chemical terms, the production of life in its animal and vegetable forms and its decay to give fossil fuels is displayed in the diagram of Fig. 6.1. Carbon dioxide and water are combined, with the help of the photosynthetic Sun's rays, i.e. the red and blue parts of the visible spectrum, to give carbohydrates consisting of molecules which are multiples of the formula $C(H_2O)$. Although green plants have been synthesizing organic materials such as carbohydrates, e.g. polysaccharides, which are the basis of the cellulose content of plants, and starch, over many millions of years, the substances have not accumulated because of their uses as foodstuffs, their microbial degradation (biodegradation) and their chemical oxidation by burning. A very small fraction of material, containing about 0.4% of the Earth's total carbon content, has been conserved as fossil fuels by avoiding the normal degradation processes by having an environment in which atmospheric oxygen was excluded. There are no microorganisms which have the ability to degrade hydrocarbons in the absence of dioxygen. Those which degrade hydrocarbons in the presence of dioxygen are responsible for the eventual clearing up of those environments which have been polluted by spillage of petroleum. If the petroleum has access to dioxygen, its microbial degradation to carbon dioxide and water is fairly rapid and eventually complete.

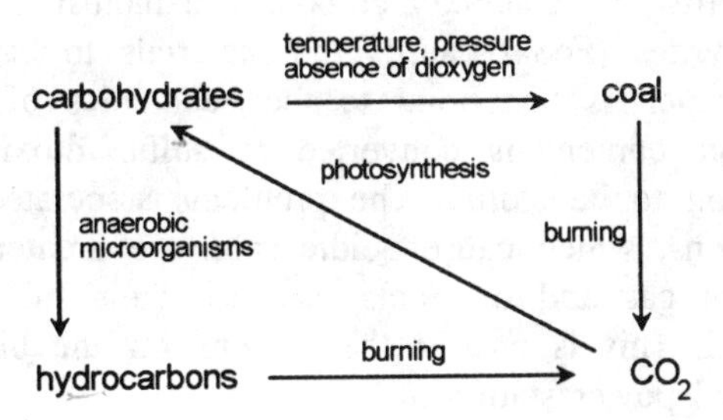

Fig. 6.1 A diagram showing the photosynthesis of carbohydrates and their conversion into coal or hydrocarbons (petroleum); the diagram also shows the re-conversion of coal and petroleum to carbon dioxide as the fuels are burned

Natural gas and petroleum are found in porous rock, mainly under high pressure which assists their extraction. They are thought to be the products from the decomposition of marine organisms. The source material has been buried by great thicknesses of sediments and, over the course of some millions of years, has been microbiologically degraded and then chemically altered under conditions of high temperature and pressure to form the mixture of hydrocarbons known collectively as petroleum. The anaerobic (absence of dioxygen), bacterially assisted, decay is shown in outline in Fig. 6.1. These liquids and gases have usually migrated from where they were first formed and have either made their way to the surface and dissipated or have been trapped in certain rock formations. The remains of the deposits which came to the surface are present as natural bitumen and tar sands.

Coal was produced from the remains of trees which were buried, so avoiding contact with dioxygen. Trees which grew in a tropical swampy environment, as they died and fell to the ground, began to decay microbiologically in the non-oxygenated water. A change in sea level caused the decaying trees to be overlain by beach sands and clays that were deposited rapidly from large river estuaries. The trees continued to decay slowly and, according to the conditions of temperature and pressure, underwent a change resulting in one or more of the end products; peat, lignite or brown coal, coal and anthracite, the latter being coal with the highest elemental carbon content. At high temperatures the organic material underwent thermal decomposition to give elementary carbon, in the form of coal, as its main constituent. This change is shown in Fig. 6.1. All these transitions are still occurring, but extremely slowly compared to the rate of the usage of the fuels, and all four materials are available as sources of energy. Peat is a porous material which has a considerable content of plant remains. Its most important use is to provide the heat and smoke which are employed in an important stage in the conversion of barley into nutritive Scottish and Irish liquid refreshments. As peat is subjected to high pressure and the natural heat of the Earth, the transformation to the other forms occurs, the carbohydrate content being converted initially to hydrocarbons and finally to elementary carbon. The inclusion of some of the beach sands and of iron pyrites (FeS_2) causes the materials to have an ash and sulfur content, neither of which is beneficial to the economics of their production and their uses. The sulfur content is converted to sulfur dioxide when the fuel is burned, the ash having to be stored. The problems associated with the control of sulfur dioxide emissions, which cause acidic rain, are greater with coal than with the alternative natural gas and oil fuels, because their sulfur contents are much less than that of coal. This is one of the reasons for the bias against the use of coal in newly designed power stations.

6.2 THE WORLD PRODUCTION OF FOSSIL FUELS

Fossil fuels are extracted from the Earth on a vast scale. The world's production of fossil fuels in 1991 amounted to 69,764,000 million cubic feet of natural gas, 24,163 million barrels of crude petroleum, and 4,469 million tonnes of coal, of which 30% was lignite. To compare these production figures it is necessary to convert them all to a common unit. The unit chosen for the comparison is the terawatt-hour, TWh. The normally used unit for electrical power is the kilowatt-hour, kWh, the lower case k (for kilo or a factor of one thousand) indicating that the unit represents the power equivalent to that used by a 1000 watt (1 kw) electric fire over a one hour period. The prefix 'tera' represents a factor of one million million (1,000,000,000,000 or 10^{12}) so that a terawatt-hour, TWh, is equal to a thousand million (10^9) kilowatt-hours. Table 6.1 shows the energy equivalents of the above fuels as terawatt-hours for their respective annual production.

For purposes of perspective, it should be noted that the world's use of nuclear fission for the generation of power in 1991 amounted to 2815 TWh. Only about 9% of oil and 14% of gas are used for electricity generation and the efficiency of their conversion into electricity is around 33% if they are burned for steam production. The high-temperature steam is used to drive the turbines which power the electricity generators. Natural gas may be burned in a gas turbine, in which the hot exhaust gases are used directly to provide the rotation required by the generator as well as producing steam for a second conventional generating stage. This is a considerably more efficient method than burning the gas to produce steam in an intermediate stage. The higher efficiency, up to 66%, of producing electricity from natural gas makes the process preferable to oil or coal burning operations. After taking these figures into consideration the contribution of nuclear power to the world's electricity generation represents about seven percent of that derived from the burning of fossil fuels. Other major sources of energy are hydroelectricity and the burning of fuel wood. Hydroelectricity production amounts to 2091 TWh per annum, the energy equivalent of the burning of fuel wood and charcoal being 6720 TWh per annum. Minor quantities of energy are supplied by geothermal sources (hot rocks) and by wind power, their annual productions being 21 TWh and 2 TWh respectively.

Table 6.1 - The energy equivalents of fossil fuels produced by the world in 1991

fuel	energy equivalent/TWh
natural gas	21100
petroleum oil	39700
coal and lignite	35800

Prices of fossil fuels are very varied and depend upon the political climate, for oil in particular, and geography, but using average figures the values of the above annual productions of natural gas, petroleum and coal are $175 billion, $448 billion and $223 billion respectively. The total value of the three fuels produced in 1991 was $846 billion.

The productions of the three kinds of fossil fuel have different geographical distributions. The distribution of natural gas production in 1991 is shown in Table 6.2, which also includes the major producer in each area and its percentage contribution to that area's production.

Table 6.2 - The distribution of world production of natural gas in 1991

area	% production	major producer/(%)
C.P.E.*	40.30	C.I.S. (93.3)
N. America	30.26	USA (80.4)
Europe	10.00	Netherlands (36.6)
Asia/Pacific	6.19	Indonesia (29.0)
Middle East	5.42	Saudi Arabia (38.5)
Latin America	4.85	Mexico (36.5)
Africa	2.98	Algeria (71.4)

* Centrally planned economies (the C.I.S. is classified in this section although its economic system is changing)

The production of natural gas by the United Kingdom represents 24.5% of the European total (i.e. 2.45% of the world total).

The distribution of world production of petroleum in 1991 is shown in Figs 6.2 and 6.3 which show the major producers in the OPEC and non-OPEC countries.

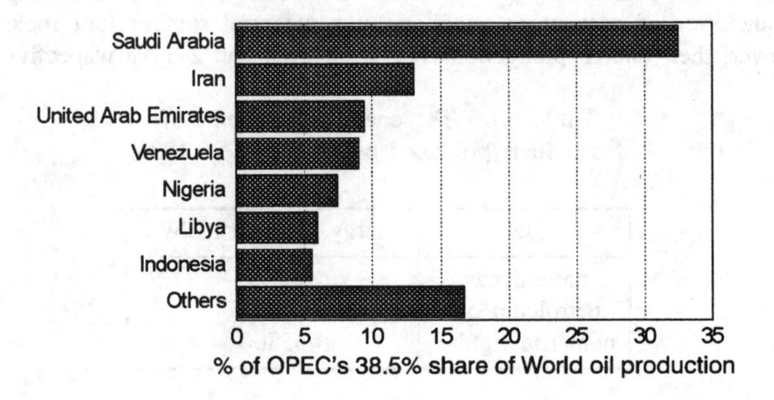

Fig. 6.2 The distribution of World production of petroleum in 1991 by OPEC countries (they produce 38.5% of the World production)

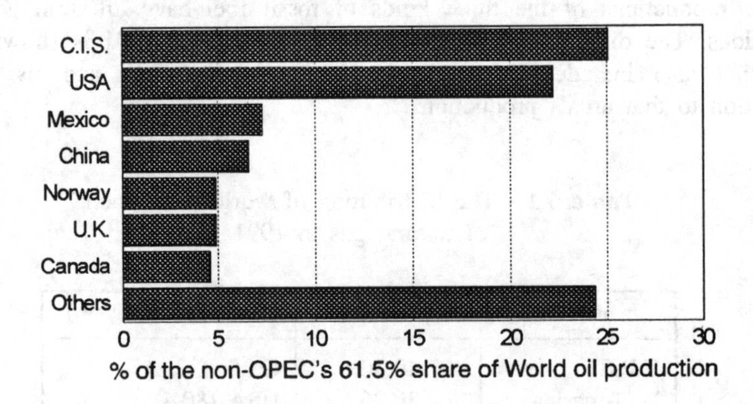

Fig. 6.3 The distribution of World production of petroleum in 1991 by non-OPEC countries (they produce 61.5% of the World production)

OPEC, which is the acronym for the Organization of Petroleum-Exporting Countries, is a powerful price-fixing group of nations and governs the rates of production of petroleum by their member countries. It accounts for 38.52% of world production, that of the non-OPEC countries being 61.48%. The price of oil as fixed by OPEC is the main determining factor of the world price, since OPEC as a retailing block is the major producer. The figures for Kuwait (an OPEC member) are included in the 'others' entry since they are unusually low (0.8%) because of the Gulf War. Normally the production from Kuwait would be in the region of 4.4% of the OPEC total.

The world production figures for coal are normally separated into those for hard coal and lignite. In 1991 the production of hard coal amounted to 3,104 million tonnes and that of lignite to 1,365 million tonnes. The distributions of world production of hard coal and lignite are shown in Figs 6.4 and 6.5 respectively.

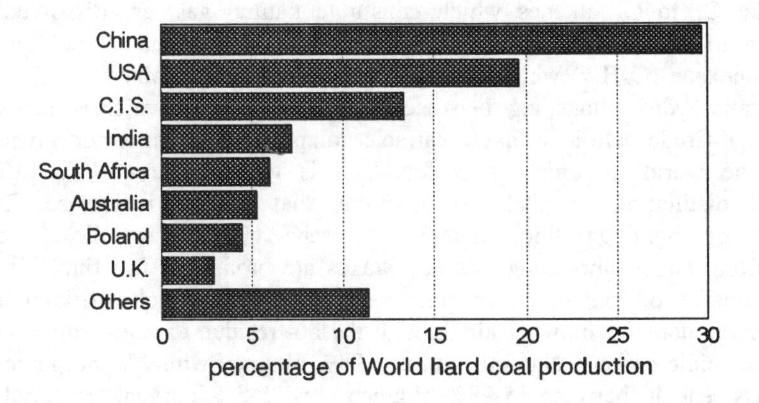

Fig 6.4 The distribution of World production of hard coal in 1991

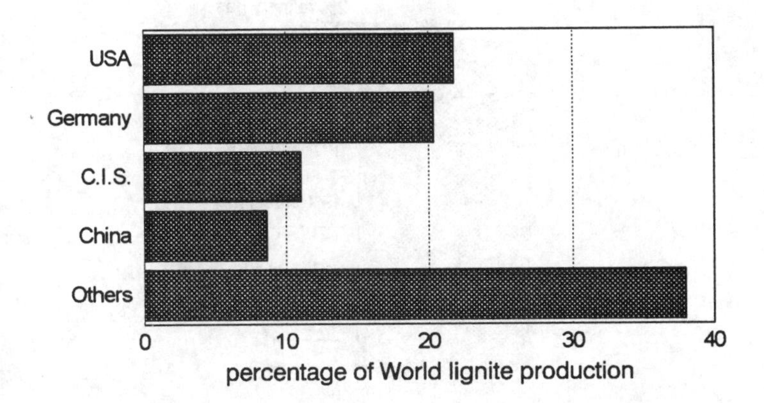

Fig 6.5 The distribution of World production of lignite in 1991

6.3 THE CHEMICAL NATURE AND USES OF FOSSIL FUELS

Details of the many ways in which carbon and hydrogen form compounds and the structures of such compounds are contained in Chapter 11. Natural gas is mainly methane (CH_4), with small amounts of ethane (C_2H_6), propane (C_3H_8), and butane (C_4H_{10}). Propane and butane are normally separated by cooling the natural gas until they condense as liquids and are used separately as liquefied gas fuels or for the production of petrochemicals.

The liquid mixture which constitutes the naturally occurring material known as petroleum consists mainly of the hydrocarbons (alkanes with the general formula, C_nH_{2n+2}) possessing between five and fifty carbon atoms. The lower members of the series, the C_1 to C_4 alkanes which constitute natural gas, are dissolved in liquid petroleum to a small extent. In addition, petroleum contains cyclic hydrocarbons, e.g. cyclohexane (C_6H_{12}) which consists of six CH_2 groups joined together in a ring, and aromatic hydrocarbons, e.g. benzene (C_6H_6) and toluene, which is methyl benzene ($C_6H_5CH_3$). Crude petroleum has a variable composition which is dependent upon the part of the world in which it is found. It is generally treated in refineries by fractional distillation to give the fractions that are of most use. The useful fractions are light gasoline, naphtha, kerosene or paraffin, diesel oil, and a residue from which lubricating oils and waxes are produced. The final black tar-like residue consists of long-chain organic compounds from which artificial bitumen is made. This is done by blowing air through the hot residue to cause some oxidation to give some acidic content. Natural bitumen (98% bitumen with 2% inorganic material), bituminous asphalt (between 15-98% bitumen plus 85-2% inorganic material) and rock asphalt (containing between 6-15% of bitumen) are found naturally, the bitumen produced by petroleum refining being regarded as artificial. Mixed with suitable minerals, in the form of sand, gravel or limestone, bitumen (artificial or natural) forms the material known as asphalt which is used for road surfacing.

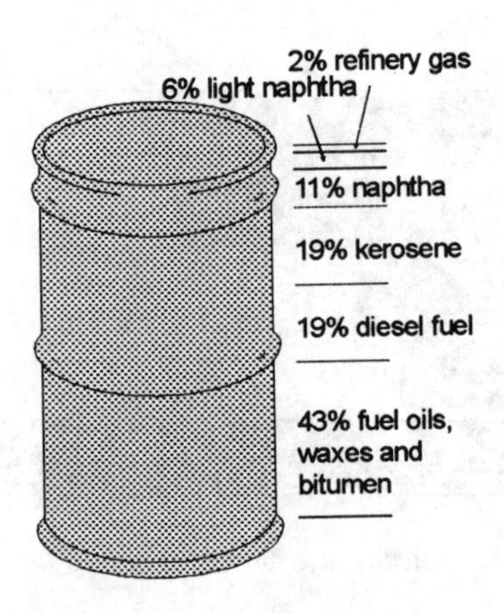

Fig. 6.6 The distribution of the fractions derivable from a typical North Sea crude oil

The distribution of the fractions derivable from a typical North Sea crude oil is shown in the diagram of Fig. 6.6. About 2% of the crude petroleum consists of refinery gas (the C_1 to C_4 hydrocarbons) which is dissolved in the oil, 6% is light gasoline, 11% is naphtha, 19% is kerosene, 19% is diesel oil, with a residue of 43% out of which waxes (~21% of the residue) and the thicker oils are extracted in subsequent processes, leaving finally the material which is converted into bitumen. Light gasoline boils between 0-70°C and contains mainly the C_5 to C_6 hydrocarbons, i.e. those hydrocarbon molecules containing 5 or 6 carbon atoms, naphtha boils between 70-140°C and contains the C_6 to C_8 hydrocarbons, kerosene boils between 140-250°C and consists of the C_9 to C_{13} hydrocarbons, diesel oil boiling between 250-350°C and containing the C_{14} to C_{25} hydrocarbons.

Until recently, tetraethyllead ($Pb(C_2H_5)_4$ in which C_2H_5 is the ethyl group) was added to all gasolines or petrols to reduce the phenomenon known as engine-knock in internal combustion engines. The knocking is caused when the final portion of the gasoline-air mixture in the engine cylinder undergoes explosive and uneven reaction. The effect of burning tetraethyl lead is to slow down the otherwise explosive reaction occurring in the engine cylinder. Other additives cause the lead to be emitted as chloride and bromide compounds, which are relatively more volatile than the oxide, and which form components of the exhaust gases of the vehicle's engine and are local atmospheric pollutants, i.e. local to the vehicle. The dispersal of the emitted lead from areas of production such as motorways is limited to about two hundred metres on either side of the road. The exhaust fumes from a car using either leaded or unleaded fuel contain several other pollutants including carbon monoxide, nitrogen oxides and particulate carbon. The toxicities of these and that of lead are discussed in chapter 5. Non-leaded petrol is more expensive to produce than the leaded variety because hydrocarbons with higher octane ratings have to be included. The price of non-leaded petrol to the motorist is kept below that of leaded petrol by a reduction in the tax rate.

When this book was written the cost of a barrel of petrol or gasoline included \$1.55 for freight, \$19.30 for the crude petroleum and \$14.80 for refining and marketing, making a total of \$35.65 (£23.30). The barrel contains 42 US gallons or 35 Imperial gallons or 159 litres. The price per litre is 14.7 pence or 22.4 cents. When the motorist buys petrol or gasoline it should be realized that any excess of price beyond these figures is made up of tax and profit.

Coal consists of elemental carbon and a large variety of carbon compounds. An elemental analysis of a typical coal gives a content of 80-82% of total carbon, 5-6% of hydrogen, 1-2% of sulfur, 1-2% of nitrogen, 3-5% of oxygen, the remainder of 5-7% being inorganic ash. There are some organic sulfur compounds in addition to the 'inorganic' sulfur content which is present as iron pyrites. Apart from its main use as a primary energy source for the production of electricity and domestic heating, coal is subjected to destructive distillation in the absence of air to prevent its burning. The products of decomposition by this process include coal gas, ammonia, tar and coke. The scrubbing, i.e. washing with water, of the gas yields the aromatic hydrocarbons, benzene and toluene. Coal gas has a typical content of 52% hydrogen gas, 32% methane, 4-9% carbon monoxide, 2% carbon dioxide, 4-5% nitrogen gas and 3-4% of ethene (ethylene) and other alkenes. It has been used for town gas but has largely been replaced by natural gas. Currently the main production of coal gas is used to heat the coke-ovens from which it is produced. Various grades of tar are produced by the distillation process which also yields light oil consisting of 60% benzene, 15% toluene and lower amounts of xylenes and naphthalene, which are all very important feedstocks for the chemical industry. The ammoniacal aqueous product, if passed into a solution of sulfuric acid, is converted into ammonium sulfate which is used by the fertilizer industry.

Natural gas is used mainly for domestic and industrial heating and for the generation of electrical power. Propane and butane are used for heating purposes where no natural gas pipeline is available. They are also used as feedstock for the production of petrochemicals. The main uses of natural gas are shown in Fig. 6.7.

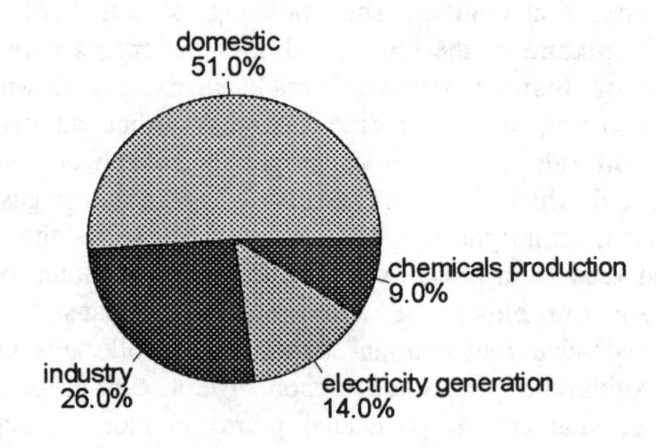

Fig. 6.7 The main uses of natural gas

The fractions of petroleum that are produced by the fractional distillation process are not always in a desirable ratio with regard to their end usage. Some of the smaller hydrocarbons are catalytically polymerized to give larger molecules. Some of the very large molecules, those in the diesel oil and residue, are subjected to catalytic cracking to give smaller molecules. Some of the light gasoline and naphtha fractions are subjected to catalytic reforming to give molecules which are more suitable for use as motor spirit. By these three processes and by the blending of the resulting fractions, fuels and oils for any purposes may be fashioned.

The immediately usable products derived from the refining of petroleum are listed as follows.

Gasolines (Light gasoline and naphtha fractions):

- Refinery gases (methane, ethane, hydrogen)
- Industrial gases (propane)
- Domestic gases (butane)
- Motor spirits (petrol, gasoline)
- Aviation fuel
- Solvents (white spirit)

Middle distillates (kerosene and lighter diesel oil):
- Kerosene (paraffin)
- Aviation turbine kerosene
- Gas oil (derv - diesel engined road vehicle - fuel, and heating oil)
- Diesel oil (marine diesel)

Fuel oil (heavier diesel oil and fraction of residue):
- Various grades of fuel oil

Residue:
- Lubricants and greases
- Waxes
- Bitumen

Between 3-6% of crude petroleum, mainly that originating in the gasoline and fuel oil fractions, is used as feedstock for the petrochemical industry. Only 15% of crude petroleum is used for purposes other than the provision of energy. The distribution of the main uses of petroleum are shown in Fig. 6.8.

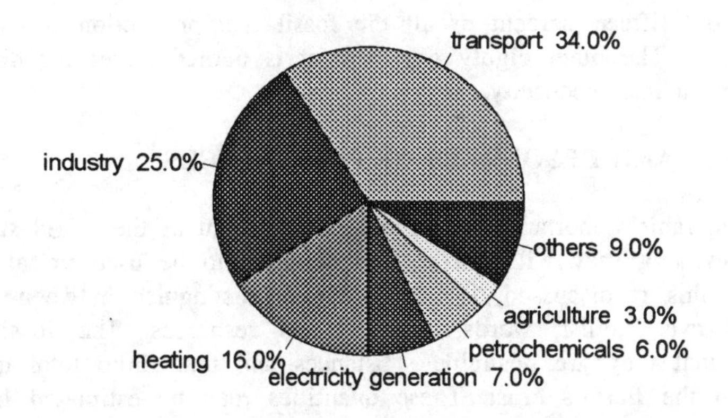

Fig. 6.8 The distribution of the main uses of petroleum

The petrochemical industry uses the feedstock from refineries to produce chemicals which are vital to agriculture and the plastics industries. Important agricultural products are fertilizers and the crop protection chemicals: herbicides, fungicides and pesticides. Kerosene burners are used in greenhouses to increase the concentration of carbon dioxide in the atmosphere, particularly when there is direct sunlight available. Under such conditions plant growth by photosynthesis is most efficient and its efficiency is dependent upon the carbon dioxide concentration in the air of the greenhouse. When the carbon dioxide content is raised by the burning of kerosene the plants grow larger and grow more quickly, offering an extra crop during the season. There are people who believe that talking to their house plants cause the plants to be more healthy. For a better health the plants require extra carbon dioxide and this is provided by the respiration of animal species. Talking to plants is not necessary for their health and casts doubt upon that of the speaker! Running on the spot or using an exercise bicycle near the plants would be even more beneficial for all concerned.

Plastics are polymeric materials made from monomers, which are the simple molecular units, produced from petroleum. They can be made in the forms of fibres, sheets or solid objects. Their names are well known and include fibres such as Nylon™, Terylene™ and Orlon™ (Acrylan™), and polythene, polypropylene, polyvinyl chloride (PVC), polystyrene, Perspex™, Formica™, and a variety of synthetic rubbers. Further treatment of polythenes and polypropylenes allow the production of a range of carbon fibres.

The long chain hydrocarbons in the residue from petroleum refining and various aromatic chemicals form the basis of the detergent industry.

About 80% of coal is used in the generation of electricity. The other major use is in the production of coke for use in the iron and steel industry, and for the production of other metals by the reduction of their oxide and sulfide ores. The other products from the conversion of coal into coke are coal gas, used to heat the coking ovens, and the chemical feedstocks for the fertilizer and plastics industries.

Only about fifteen percent of all the fossil fuel production is used for non-energy purposes. The other eighty five percent is burned either for direct heating or for conversion into electricity.

6.4 RESERVES AND RESOURCES OF FOSSIL FUELS

It is of considerable importance to appreciate the extent of the world supply of the fuels and how long they will last if they continue to be used up at the present rates. Before this is discussed, it is essential to distinguish between the specific meanings of two English words; reserves and resources. The fossil fuels are resources in that they are desirable substances and that finite total quantities of them exist in the Earth's crust. These quantities may be estimated from current knowledge but are likely to change as knowledge increases. The quoted amount of a resource is the known total of the material, whether it is feasible or not to extract it from the earth. Exploration and drilling give information about the quantities of the fuels which are available. The amounts which are available and which have been proved to be removable from the earth, at current prices and with current technology, are called reserves. Estimates of both the resources and the reserves of gas, petroleum and coal, vary with time as more exploration is carried out.

Finds of natural gas and petroleum in recent years have increased estimates of the resources and the extractable reserves. The most recent estimates (1989) of the number of years before gas and petroleum supplies run out are about fifty nine years and forty six years respectively. The reserves of coal are larger than those of natural gas and petroleum and should last for about another 210 years (1992 figures). As an example of the differences between reserves and resources the figures for coal are 853,000 million tonnes and 8,330,000 million tonnes respectively, the smaller figure representing the coal which can be economically recovered, i.e. only about 10% of the known deposits of the fuel. The figures are necessarily approximate and are based upon the best available knowledge of the extent of the reserves, and it is assumed that there will be no changes in the rates of use of the resources. As stocks decrease there will be greater efforts applied to the extraction of resources which are currently uneconomic. World politics and economics may be determined in future by the distribution of the resources, in particular those of natural gas and petroleum. The members of OPEC possess 66% of the world's reserves of petroleum and 32% of the world's reserves of natural gas. The comparative figures for the C.I.S. are 12% of petroleum and 40% of natural gas. These are followed by the USA with 5.7% and 9.3%, other countries being very much lacking in the two resources. The continent of Europe has only 4.3% of the reserves of petroleum and 6.5% of the reserves of natural gas. The North Sea oil wells have sufficient proven reserves to last for around twenty years, the natural gas reserves being sufficient for between thirteen years (U.K.) and seventy seven years (Norway).

The countries and regions of the world may be assigned to one or other of five main groups depending upon their economies and their usage and ownership of the energy minerals. The first group are the members of OPEC, who own much of the world's reserves of petroleum and about a third of the natural gas. They are large producers and exporters, but consume little themselves. They consist of largely non-industrialized countries dependent upon their exports for the maintenance of their economies. The second group consists of the industrialized countries with unplanned economies ('the West') which are heavy users of energy but are dependent upon imports to maintain their systems. A third group consists of the recently formed Commonwealth of Independent States. The C.I.S. is currently undergoing changes in its economy but is self sufficient in energy minerals and is a heavy user. The fourth group consists of the countries with centrally planned economies (mainly China) which are rich in coal but have very little oil and gas reserves and are comparatively low users of energy. The fifth group contains all the other countries ('the Third World') who are poorly supplied with energy minerals but are also low users. They suffer from the high world prices of imports of energy and consumer goods which is a very large factor in the retardation of their development. The influence of chemistry upon the economic and social development of the countries of the world is very great and has been largely ignored by historians.

6.5 RELATIVE CALORIFIC VALUES OF FOSSIL FUELS

The energy content of a fuel is quoted as a calorific value; the amount of heat energy which is released when a certain unit of weight or volume of the fuel undergoes complete combustion. To be able to compare such different fuels as natural gas, fuel oil and coal it is necessary to use the same units for their calorific values. The unit chosen is the number of kilowatt-hours which is contained by twelve grams, i.e. one mole, of contained carbon. The values for the fossil fuels and some model substances are given in Table 6.3.

Table 6.3 - Calorific values for fossil fuels and some model substances

substance	calorific value kWh per mole of carbon
graphite	0.109
coal	0.109
methane	0.247
natural gas	0.245
hexadecane, $C_{16}H_{34}$	0.184
fuel oil	0.184

The calorific values for coal and graphite, which is pure carbon, natural gas and methane, and fuel oil and hexadecane, show that the fossil fuels are very similar in heat energy content to the substances chosen for comparison. The cost per kWh for domestic purposes is given for the three fuels in Table 6.4, together with the costs of using electricity at day and off-peak rates.

Table 6.4 - Domestic costs (U.K.) of fossil fuels compared with those of electricity

fuel	U.K. cost p/kWh
natural gas	1.5
fuel oil	1.4
coal	1.2
electricity	7.5 (day)
	2.8 (off-peak)

The comparison does not take into account the very different capital costs associated with the installations required for the conversion of the various fuels into heat, nor does it take into account the efficiencies of such conversions. The considerably higher cost of electricity arises because of its being a secondary fuel, dependent upon burning of one or other of the fossil fuels for its production and demonstrating the low efficiencies and higher expense of such processes. There are other factors built into the pricing systems of the natural gas and electricity industries such as standing charges which are not included in the figures in Table 6.4.

6.6 ALTERNATIVE ENERGY SOURCES

Alternative sources of energy to the use of fossil fuels are:

(i) nuclear fission (discussed in Chapter 3)
(ii) nuclear fusion (discussed in Chapter 3)
(iii) solar energy, which may be converted into electrical power by photocells or used for heating water or concentrated by mirrors for heating solar furnaces,
(iv) wind and wave power, and
(v) heat extracted from hot rocks, which is geothermal energy.

The conversion of nuclear fission energy into electricity is well developed. Research and development of the other sources of energy are only in their early stages. The use of nuclear fusion offers the possibility of 'cleaner' energy than that from nuclear fission, but has not been developed beyond the research stage yet. There is no doubt that viable alternatives to the use of fossil fuels will have to be developed within the next twenty five years, especially if the economies of the world are to be maintained. The fossil fuels will remain as the main source of feedstock substances for the chemical industry and should be conserved for that purpose. Eventually the chemical industry will have to use freshly grown vegetable sources other than the fossil fuels for the continuing manufacture of essential substances and materials. When it becomes necessary, fuels will be manufactured from vegetable materials. Vigorous research into the production of oil from rape-seed is taking place with encouraging results from pilot-plant scale operations.

Solar power represents a very large resource of energy. The energy which actually penetrates the atmosphere and is incident upon the Earth's surface has an intensity of 155 watts per square metre for surface which is in the equatorial regions, a better average figure being around 100 watts per square metre. If an area of 25 square kilometres (10 square miles) were to be used for solar energy collection, it would receive 2500 megawatts watts of power. The inefficiency of converting the solar radiation into electricity and the restictions of the weather and the number of daylight hours would reduce the output of the area to about 1000 megawatts. Around ten thousand of these solar power collection areas would be needed to supply the energy requirements of the current population of the Earth. The total area covered by this number of power stations would be a quarter of a million million square kilometres (one hundred thousand square miles), an area which is equivalent to that of the state of Utah or twice that of England.

Various physical and chemical methods of capturing solar energy are being investigated. These range from those individual houses which use the Sun's rays to heat water which is then used for heating purposes to solar panels consisting of semi-conductor materials which allow the direct conversion of the Sun's light into electrical energy. Such technology is available at the present time but the high cost, together with the low conversion efficiency, makes the process uneconomic. A large area is needed for solar radiation collection by photocells, the efficiency of the process being hindered by the effluent from the bird population.

The derivation of energy from wind power is under investigation in various parts of the world. The wind parks or wind farms take up a large amount of space, the individual units being expensive, using much steel for construction and copper in the generators, to build and to run. The claim that wind powered generators have no environmental impact must be tempered by the consideration of the impact on the environment of the production of the necessary steel and copper. Although it would not be practical, power equivalent to that currently produced by conventional means in the United Kingdom would require the erection of sixteen million wind powered generators. There are no plans to use wind to such an extent, but the experimental wind farms are already producing protests from people in their locality. The output from wind powered generators, being dependent upon the weather, is irregular and difficult to control. Even if wind power were to be used to produce electricity there would have to be a suitable back-up supply for windless conditions. To duplicate supply methods is expensive and effectively rules out wind power as a serious contributor to electricity supplies. The derivation of energy from wave and tidal power is only at a very early stage of development. Geothermal power requires very expensive deep drilling into suitable hot rocks and currently is showing little potential for development in the near future.

7

Water

Molecular structure of water. Properties and structure of liquid water and ice. Rain. Snow. Rivers. Oceans. Acidic rain. Water purity. Ion exchangers. Hardness of water. Soaps and detergents. Sea bed minerals. Composition of sea water. Salt, caustic soda, chlorine, hydrogen and soda ash. Baking powder.

7.1 MOLECULAR STRUCTURE AND PROPERTIES OF WATER

The water molecule is the most abundant molecule on Earth. It has a triangular structure with the two hydrogen atoms covalently bonded to the central oxygen atom as shown in Fig. 7.1. The O-H bond length is 96 picometres, the bond angle is 104.5°.

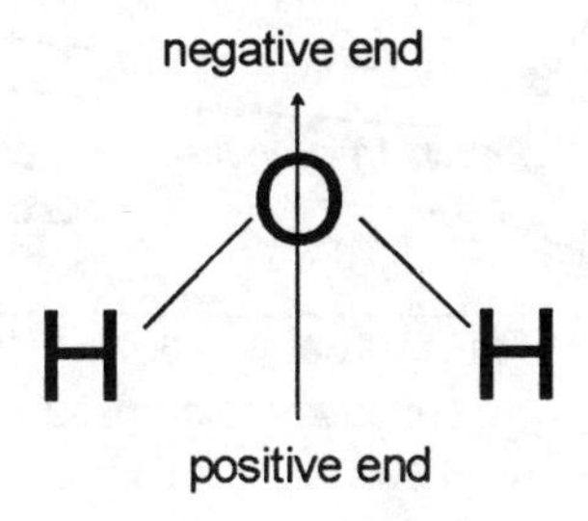

Fig. 7.1 The structure of the water molecule showing its polarity

The electron attracting properties of hydrogen and oxygen are sufficiently different to make the O-H bonds quite polar, the oxygen end of the molecule being slightly negative with respect to the positive hydrogen end. The two elements do not differ sufficiently to allow an ionic bond to be formed; the best description of the bonding is that it is polar covalency. The oxygen atom possesses the greater share of the available electrons. The polarity of the molecule is the main factor responsible for the almost unique properties of the liquid and solid states of the substance. At the standard pressure of one atmosphere, water melts at 0°C and boils at 100°C, these temperatures being the fixed points on the Celsius scale.

As a broad generalization it appears that the physical constants, melting and boiling points, of molecular substances with similar chemical composition, i.e. comparable compounds of elements belonging to the same Group of the periodic table, are roughly proportional to their relative molar masses, RMM. This also applies to the monatomic Group 18 elements, helium, neon, argon, krypton and xenon, and is generally understood in terms of the interatomic and intermolecular forces being fairly constant in most cases. In order to melt a solid, the intermolecular forces have to be overcome and the molecules of the liquid state have to acquire sufficient energy to move around independently of their neighbours. The latter factor is dependent upon mass, making heavier molecules more difficult to convert to the liquid state. Similar considerations apply to the variations in boiling points as the RMM of the substances increase. Only if there is a particularly strong force operating are there exceptions to these general trends. The boiling points of three series of substances are plotted against their individual RMM values in the graph shown in Fig. 7.2.

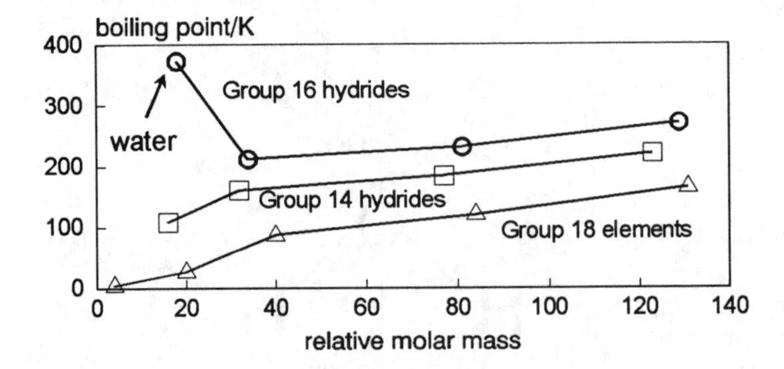

Fig. 7.2 A graph of the boiling points of the Group 18 elements (He, Ne, Ar, Kr & Xe), the Group 14 (C, Si, Ge & Sn) and Group 16 (O, S, Se & Te) hydrides (formulae; EH_4 and EH_2 respectively, E representing the Group 14 or 16 element) against their RMM values; the b.p. of water is anomalous

The boiling points of the Group 18 elements increase in a roughly linear manner with increasing RMM value. Those of the series of hydrides with the formula, MH_4, where M = carbon, silicon, germanium, tin or lead (Group 14 elements) also show an approximately linear relationship with their RMM values.

The values for the three hydrides with the formula, MH_2, where M = sulfur, selenium or tellurium (Group 16 elements), show an almost linear relationship; but the value for the lightest hydride of the group, i.e. water, is exceptionally high (100°C, 373.15 K) for its low RMM value of 18. The values for the heavier hydrides of the Group 16 series indicate that the boiling point of water would be around -75°C if the same kind of forces operated between water molecules as those between the heavier molecules. The extra intermolecular force which operates between molecules of water is known as hydrogen bonding and is noticeable in compounds of hydrogen with very electronegative elements where the hydrogen-element bonds are very polar. Extra energy has to be supplied to water to cause the solid-to-liquid and liquid-to-gas changes in order to break down the hydrogen bonds. In the absence of hydrogen bonding the world would be very different as all the water would be in its gaseous state!

Hydrogen bonding has a large influence upon the crystal structure of solid water (ice) causing it to consist of an open framework with each oxygen atom having four hydrogen atom neighbours, two of which belong formally to two other water molecules. The solid state of water is remarkably complex. There are two crystal forms which can exist at atmospheric pressure, but at higher applied pressures of between 2 to 25 atmospheres at least six other crystalline structures have been identified. The normal form of ice, as it is formed naturally, is a hexagonal structure, as demonstrated by the shape of snowflakes, similar to that of the silica mineral known as tridymite. Ice has a density of 0.92 grams per cubic centimetre (cc) at 0°C and floats on liquid water (density = 1 gram per cc) at the same temperature. This unusual property is because of the greater degree of hydrogen bonding in the solid form. Melting is associated with a partial collapse of the more open hydrogen bonded structure which allows the individual molecules to be closer to their neighbours than they are in the solid state. The opposite is generally observed when substances are heated. Their constituent molecules acquire energy and need more space in which to vibrate more energetically. If water had the normal property of being more dense in the solid state aquatic life would be difficult, if not impossible, in the colder regions of the oceans. The *Titanic* would still be sailing today because ice bergs would be situated on the sea bed. In fact, ships would not be necessary because water would be a gas without hydrogen bonding. Life on Earth owes a great deal to the hydrogen bond because living matter, whether animal or vegetable, has liquid water as its primary constituent.

Another property of liquid water which is vital to life, and many other features of the Earth, is its action as a solvent. The solvent action of water is related to its polar nature which allows strong interactions with positive or negative ions and with molecules which possess polar groupings of atoms, e.g. sugar, the sucrose molecule (its structure is shown in Fig. 9.3) which has eight hydroxyl (OH) groups and is extremely soluble in water. Molecules which are non-polar, such as hydrocarbons, e.g. oil and petrol/gasoline, do not dissolve in water to any appreciable extent.

7.2 THE DISTRIBUTION OF WATER ON EARTH

The total amount of water on the Earth has been estimated to be around 1.463 million million million tonnes (1.463 x 10^{18} tonnes), 95.7% constituting the oceans, and 4.1% as ground water contained in soil, underground streams, and aquifers, which are porous and cracked rocks. The remaining 0.2% is contained in polar and mountain-top ice, lakes, swamps and rivers, together with some in the gaseous form as water vapour and water droplets in the atmosphere. The oceans lose about 350 teratonnes (one teratonne - 1 Ttonne - is a million million tonnes, 10^{12} tonnes) of water annually by evaporation, the evaporation from land based sources amounting to 70 Ttonnes. Of the total evaporation of 420 Ttonnes, only 14 Ttonnes are present in the atmosphere at any given time. On average, the evaporation rate, equal to the rate of precipitation as rain, hail and snow, is calculated as $420 \div 365 = 1.15$ Ttonnes per day. Because there are 14 Ttonnes in the atmosphere on any given day, the average residence time of water in the atmosphere is calculated as $14 \div 1.15 = 12.2$ days. The precipitated water falls onto the Earth's surface as rain, hail and snow, the land area, which represents 29% of the total surface, receiving 23.8% of it. This means that there is a transfer of 30 Ttonnes of water from the oceans to the land in one year. This amount of water replenishes the ground water, aquifers, glaciers, snowfields, lakes, swamps and rivers, the latter duly delivering 30 Ttonnes per year back into the oceans. The distribution of water on the Earth and its annual transfer rates between the land, the oceans and the atmosphere are shown in Fig. 7.3.

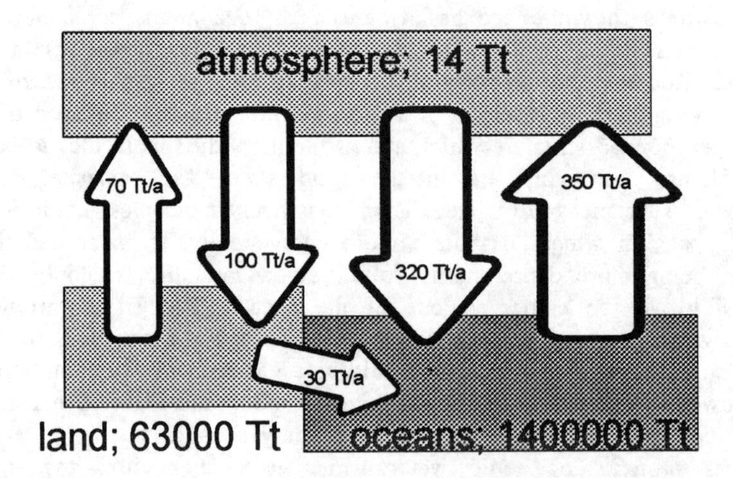

Fig. 7.3 The water cycle in units of teratonnes (Tt) for the contents of the atmosphere, the land and the oceans, and in units of teratonnes per annum (Tt/a) for the transfers of water between them by evaporation, precipitation (rain and snow) and river flow

7.3 ACIDIC RAIN

When clouds become sufficiently cool, raindrops are formed which then fall to the ground either as rain, or as hail or snow if the conditions are appropriate. Whilst falling gently to Earth the rain performs a cleansing action including the removal of dust particles and the dissolution of any sulfur dioxide in the air. The water droplets also dissolve some carbon dioxide. Gases such as dioxygen and dinitrogen (the major constituents of air) dissolve in the raindrops only to a slight extent because of their non-polar nature. Carbon dioxide is also a nonpolar molecule but reacts with water to give carbonic acid, H_2CO_3. This is a weak acid which to a slight extent undergoes dissociation to give aquated hydrogen ions and bicarbonate ions:

$$CO_2(g) + H_2O(l) \rightleftharpoons H_2CO_3(aq) \rightleftharpoons H^+(aq) + HCO_3^-(aq)$$

the combination of the reaction with water and the dissociation of the product causing carbon dioxide to have an appreciable solubility in water. At 15°C the solubility of atmospheric carbon dioxide in water is sufficient to produce a solution with a pH value of 5.6 which is the normal value for rainwater. Unlike carbonic acid, the acids produced from sulfur dioxide in the atmosphere, sulfur(IV) acid (sulfurous acid, H_2SO_3) and sulfur(VI) acid (sulfuric acid, H_2SO_4), are strongly dissociated into aquated hydrogen ions and their counter ions. Sulfur(IV) acid is very rapidly converted to sulfur(VI) acid by reaction with dioxygen in the presence of radiation or by reaction with ozone. The small water droplets of which clouds are composed could contain a sufficiently large concentration of sulfur(VI) acid to cause the pH value of the eventual rainwater to be as low as 4.0 which is the same acidity as apple juice. The distribution of sulfur dioxide and its eventual product by winds in the atmosphere has the effect of producing acidic rain in areas sometimes far away from the point of the emission. It is possible for the substances to travel as much as 4000 km before they are removed by precipitation, although it is much more common for their removal to take place within 200 km. Some of the atmospheric sulfur dioxide and its products are removed by dry reactions with limestone, whether it be in the form of outcrops, cliffs, statues or buildings. The acidic clouds which are formed mainly in the heavily industrialized areas of the world, i.e. the eastern side of North America and the whole of western Europe, deposit their acidity in those same areas. The rest of the world seems not to suffer from the problem. Where acidic rain is precipitated it can cause considerable damage to crops and to natural vegetation. There are three main ways in which damage to vegetation can occur. One is the direct damage to green leaves by the falling acidic rain which may dissolve essential ions such as the magnesium ion which is required for photosynthesis to occur. Leaves which are deficient in magnesium turn brown and fall off. The second effect is the leaching of essential nutrients from the soil. The third is the solubilization of toxic elements such as aluminium, copper, lead and zinc from the mineral content of the soil.

Acidic rain poses a serious threat to aquatic life in lakes because of the decrease in pH value of the water and the solubilization of toxic elements. The acidity can be counteracted by lime treatment, but that is a temporary measure and has to be continually repeated to have a significant effect.

Many countries have Clean Air Acts of legislation which are applied fairly rigorously and have caused local improvements in air quality, but the acidic rain problem is one in which the cause and effect often lie in different countries. The scale of the problem is very large when the amounts of the emissions of sulfur dioxide are considered. There are natural emissions from volcanoes which account for about 18% of the total sulfur dioxide entering the atmosphere. The natural decomposition of materials containing proteins, which contain sulfur, accounts for around 26% of the total, any hydrogen sulfide being easily oxidized to sulfur dioxide, leaving 55% which has its origins in the burning of domestic and industrial fuels and in the smelting of sulfide mineral ores for metal production. The 55% of emissions, over which there is some control, represents 90 million tonnes of sulfur dioxide per annum. Various chemical methods are available to prevent sulfur dioxide from entering the atmosphere. It is possible to minimize the sulfur in coal by the bacterial oxidation of its pyrite content but that leaves the organic sulfur content, that present as constituents of organic compounds, to form sulfur dioxide when the coal is burned. It is possible to use bacterial oxidation as an alternative to the roasting of sulfide ores in metal extraction processes. It is also possible to leave the sulfur removal until after the coal has been burned and after the sulfide ores have been roasted. The flue gases may be de-sulfurized by reacting the acidic sulfur dioxide with a suitable basic material. Possible basic substances have to be reasonably cheap and to react reasonably quickly with sulfur dioxide. The use of limestone, i.e. calcium carbonate, or lime, i.e. calcium oxide which is produced by heating limestone, causes the conversion of the sulfur dioxide into calcium sulfate with the formula, $CaSO_4.2H_2O$, which is the same as the natural mineral known as gypsum. The product can be dried and used in plaster, plaster board manufacture or in road building. If dolomite ($MgCO_3.CaCO_3$) is used, it is transformed into magnesium sulfate from which the sulfur dioxide can be regenerated by heating the dried substance and converted into sulfuric acid. Both products of flue gas de-sulfurization processes, i.e. gypsum and sulfuric acid, are more economically produced by other processes thus making the treatment of flue gases expensive. It is cheaper to dump the gypsum produced than to transport it to a manufacturer of plasterboard. Another factor to be considered is that the de-sulfurization process would use about two hundred thousand tonnes of limestone annually and that would have to be mined. A typical coal-fired power station would produce around one million tonnes of wet gypsum sludge per annum, the equivalent of two hundred thousand tonnes of concentrated sulfuric acid. Oil-fired power stations share the problem but those which burn natural gas do not.

7.4 THE PURIFICATION OF WATER

Of the world's population of some 5.3 billion people, some 1.2 billion lack safe drinking water and around 25,000 of them die daily from water borne diseases. The water used for drinking and by industry in the developed parts of the world is treated with a coagulant such as aluminium hydroxide, produced by adding aluminium sulfate (alum). The aluminium hydroxide precipitate has a great attraction for fine particles including colloidal minerals, pollen and bacteria. After filtration, the water is softened by ion exchange with natural zeolites, and then it is sterilized or disinfected by chlorination or by treatment with ozone. The filtration removes solids and allows oxidation of the majority of organic pollutants to carbon dioxide. Chlorination renders the water sterile. The only pollutant to escape the purification processes is nitrate ion, NO_3^-, which is continually being leached out of the soil which supports vegetation, particularly that used for agriculture. The rivers of the world have a measurable nitrate ion content, some data for which are shown in Fig. 7.4.

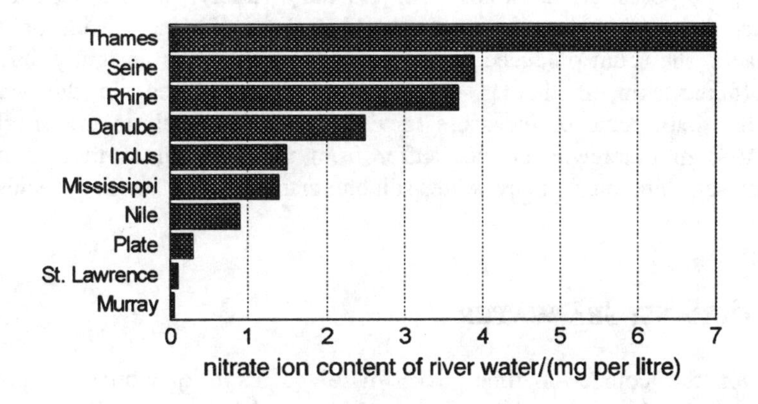

Fig. 7.4 The nitrate ion content of some rivers

The safe level of nitrate ion content of drinking water is recommended to be below 50 milligrams per litre by the World Health Organization. Even though there is extensive nitrate ion leaching occurring, it would appear that no natural waters contain a sufficient concentration to endanger health of the drinkers. People who use domestic filtration units to purify their drinking water even further are running no risks whatsoever to their health. The ion-exchange resins in the units are designed to remove positive ions by replacing them with aquated hydrogen ions and to remove negative ions by replacing them with hydroxide ions. The hydrogen ions react with the hydroxide ions to give water. Equations expressing the reactions of water purification are:

$$R_a\text{-H} + M^+ \longrightarrow R_a\text{-M} + H^+$$

$$R_b\text{-OH} + A^- \longrightarrow R_b\text{-A} + OH^-$$

$$H^+ + OH^- \longrightarrow H_2O$$

R_a and R_b representing the remainder of the acid and base exchange resin structures respectively, M being any soluble metal and A^- representing any negative ion. The resins in the purification units remove the ions of calcium and magnesium which give the water its natural 'hardness' and which, if left in the water, allow the formation of unwanted scale on the electrical heaters of kettles and immersion heaters.

The hardness of water is classified as temporary and permanent. Hardness occurs because of the presence of dissolved calcium and magnesium salts (chlorides and sulfates) and carbon dioxide as the bicarbonate ion, HCO_3^-. When the water is heated, a precipitate of calcium and magnesium carbonates is formed, which removes the temporary hardness. Any remaining calcium and magnesium content is responsible for the permanent hardness which is usually apparent when the water is used with soap for cleaning purposes. If the water is of the hard variety, the soap reacts with the calcium and magnesium ions to give insoluble salts of the fatty acids of which the soap is made. The scum produced under such circumstances is unsightly, but is easily washed off the skin, if that is the surface being cleansed. In the washing of clothes with soap, some of the scum inevitably attaches itself to the articles being cleaned. Modern detergents do not suffer from the problem as their calcium and magnesium salts are much more water-soluble than those of the fatty acids used in soaps.

7.5 THE OCEANS; SEA WATER

Rivers, over the course of time, have dissolved large quantities of the soluble salts from the Earth causing the oceans to be very rich in some elements. The flow of rivers causes the movement of great quantities of solid material which acts in an abrasive manner to erode further amounts of solids. Solid materials are deposited at the mouths of rivers, particularly in those which have deltas, and in the nearby seas. Other material is constantly being ejected from the ocean beds by volcanic and more gentle processes known as oozing, which contributes to the very large resources of minerals in those regions. Sharks teeth, fish bones, shells and small pieces of rock form nucleating sites for what are called manganese nodules. These are objects around the size of a potato and are rich in manganese (35% hence the name), with considerable concentrations (0.2-2.5%) of metals which include copper, nickel and cobalt. Estimates of their abundance indicate that there may be around 1500 million million tonnes of these nodules lying on the bed of the Pacific Ocean. They are being produced at something like 10 million tonnes per annum. Recovery of metals from these sources has not yet begun on a commercial scale.

The main constituents of sea water, those with abundances of more than one milligram per litre, are given in Table 7.1. There are considerable quantities of at least forty five other elements present in sea water with concentrations from one milligram per litre down to as little as ten nanograms per litre. These are small concentrations but when the mass of the oceans (1.4×10^{18} tonnes) is considered amount to vast quantities of such elements.

Table 7.1 - The concentrations of the main constituent elements dissolved in sea water

element	concentration milligrams per litre
chlorine	19000
sodium	10500
magnesium	1350
sulfur	885
calcium	400
potassium	380
bromine	65
carbon	28
strontium	8
boron	5
silicon	3
fluorine	1

The chlorine, sulfur, bromine, carbon, boron, silicon and fluorine are present as the negative ions, chloride, sulfate, bromide, carbonate, borate, silicate and fluoride respectively. The metallic elements are present as their positive ions, Na^+, Mg^{2+}, Ca^{2+}, K^+ and Sr^{2+}. One litre of sea water, when slowly evaporated to dryness, yields around 42.8 grams of solid salts, the main constituents being sodium chloride (NaCl, common salt, 58.9%), magnesium chloride ($MgCl_2.6H_2O$, 26.1%), sodium sulfate ($Na_2SO_4.10H_2O$, 9.8%), calcium sulfate ($CaSO_4$, 3.2%) and potassium sulfate (K_2SO_4, 2%). The distribution of chlorides and sulfates arises because of their relative solubilities, the less soluble sulfates being formed before the more soluble chlorides crystallize.

7.6 SALT, SODIUM CHLORIDE (NaCl)

The use of salt to enhance the flavour of food is well known. The chlor-alkali industry uses salt for the electrolytic production of caustic soda (sodium hydroxide, NaOH) and gaseous dichlorine, Cl_2. One of the methods used for this process is to use a liquid mercury cathode (negative electrode) with titanium anodes (positive electrodes) with brine as the electrolyte solution. Passage of electrical current through the cell causes the positive sodium ions to be discharged (neutralized) at the mercury cathode where it forms an alloy or amalgam with the mercury. The negative chloride ions are discharged at the titanium anodes and form dichlorine gas which is collected and stored as the liquid in cylinders. The mercury cathode is continuously being changed, fresh mercury flowing into the cell as the sodium amalgam flows out. The sodium-rich mercury is treated with water which converts the sodium into sodium hydroxide solution:

$$2Na\text{-}Hg + 2H_2O \longrightarrow 2Hg + 2NaOH + H_2$$

the dihydrogen being collected and stored in cylinders at high pressure (2000 atmospheres). The use of mercury cells is being phased out in favour of alternatives which make a smaller environmental impact. There is an emotional pressure-group of people who show great zeal, but little understanding, about the inextricably linked manufacture of chlorine and alkali ('*you can't have one without the other*', as the song goes) in demanding the cessation of chlorine production. If that were to happen, one consequence would be a disastrous increase in the incidence of water-borne diseases. There would be a greater dependence upon soda-ash to substitute for sodium hydroxide which would lead to a greater use of limestone and/or a faster use of the world's trona reserves, as discussed below.

Sodium hydroxide or caustic soda is used for the manufacture of many chemical substances, in the wood pulp and paper industries, for the conversion of bauxite to alumina, and for the manufacture of soap. Chlorine has many important uses, the production of the plastic, polyvinyl chloride (PVC), and the bleaching of wood pulp for paper manufacture being the two most significant. The hydrogen produced, although sodium hydroxide production is not the main source of the gas, is used for the manufacture of ammonia by the Haber process and has many uses in the chemical industry, in welding and as the fuel for the main engines of the Space Shuttle. Hydrogen can be used to drive suitably converted car engines. The combustion product of hydrogen is pure water and in the *hydrogen economy*, currently the subject of much debate, atmospheric pollution from the car would be zero.

The world production of salt in 1990 amounted to 183.6 million tonnes, its distribution being shown in Fig. 7.5. The percentages of the major uses of the world's supplies are shown in Fig. 7.6.

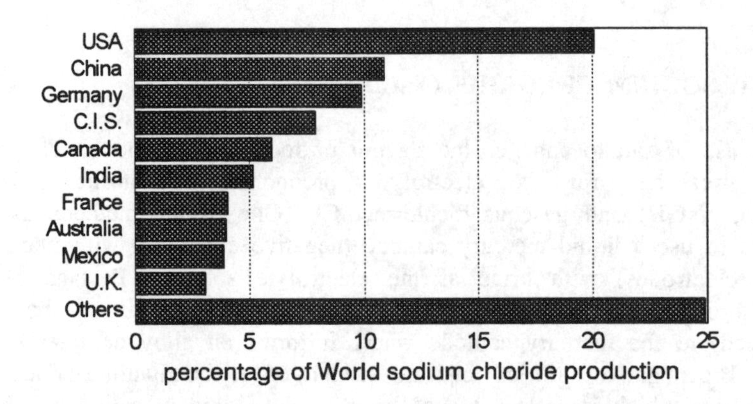

Fig. 7.5 The distribution of World production of salt (sodium chloride) in 1990

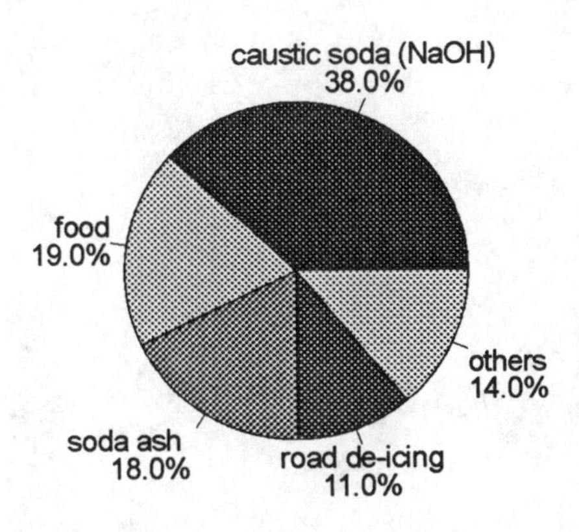

Fig. 7.6 The major uses of salt (sodium chloride)

Soda ash is the commercial name for sodium carbonate and is produced from salt by the Solvay process in which limestone ($CaCO_3$) and sodium chloride are used. The overall reaction is represented by the equation:

$$CaCO_3 + 2NaCl \longrightarrow Na_2CO_3 + CaCl_2$$

although the reaction goes spontaneously in the opposite direction. That the substances on the left hand side of the above equation do not normally react together is clear to anyone who has observed waves in the sea (a dilute solution of sodium chloride) breaking against the white chalk cliffs of Kent and Sussex in Southern England. The two potential reactants are shown in contact with each other in the photograph shown in Fig. 7.7. A mixture of solutions of sodium carbonate and calcium chloride produces a precipitate of chalk, this being one reaction which is used to produce pure precipitated chalk. The Solvay process is a very good example of how chemists, by using energy, are able to reverse a spontaneous process. The first stage in the Solvay process is to heat limestone to give lime and carbon dioxide:

$$CaCO_3(s) \xrightarrow{\text{heat}} CaO(s) + CO_2(g)$$

Saturated sodium chloride solution (brine) is then treated with ammonia and carbon dioxide. The ammonia reacts with the aqueous carbon dioxide to give ammonium hydrogen carbonate (ammonium bicarbonate) solution:

$$NH_3 + CO_2 + H_2O \longrightarrow NH_4HCO_3$$

Fig. 7.7 A photograph of the Seven Sisters chalk cliffs near Seaford in Sussex, England, in contact with the sea (sodium chloride solution). The two substances do not react with each other spontaneously [Photo: J.J.Barrett]

Ammonium hydrogen carbonate then reacts with the sodium chloride:

$$NH_4HCO_3 + NaCl \longrightarrow NaHCO_3(s) + NH_4Cl$$

giving ammonium chloride solution and solid sodium hydrogen carbonate which is insoluble in the brine. The solid is separated from the solution by filtration and heated to 150°C at which temperature it decomposes to give the desired product, sodium carbonate, and regenerates carbon dioxide which is re-used:

$$2NaHCO_3 \longrightarrow Na_2CO_3 + CO_2 + H_2O$$

The lime produced in the generation of carbon dioxide is then used to regenerate ammonia from the ammonium chloride solution:

$$CaO + 2NH_4Cl \longrightarrow CaCl_2 + 2NH_3 + H_2O$$

leaving calcium chloride as the only waste product, no major uses having yet been found for it.

The deposits of the mineral, trona ($Na_2CO_3.NaHCO_3.2H_2O$), found in the USA at Trona in Southern California, Kenya and Egypt are now used to produce a high proportion of the world's supply of soda ash. The effect of heating trona is to drive off the water and some carbon dioxide is produced:

$$2Na_2CO_3.NaHCO_3.2H_2O \longrightarrow 3Na_2CO_3 + CO_2 + 5H_2O$$

The major uses of soda ash are in the manufacture of glass, sodium phosphates from naturally occurring calcium phosphates, for water softening purposes and use in cleaning materials.

Sodium hydrogen carbonate, better known as sodium bicarbonate or bicarbonate of soda, is used in cooking as a component of baking powder, the other component being sodium hydrogen phosphate. There is no reaction between these components if the mixture is dry but reaction does occur when water is present, as is the case in the preparation of dough and pastry. The aquated hydrogen ions furnished by the acid phosphate ion react with the hydrogen carbonate ions to produce water and gaseous carbon dioxide:

$$H^+ + HCO_3^- \longrightarrow H_2O + CO_2$$

the carbon dioxide causing the required rising of the preparation.

8

Some important substances and materials

Sand. Limestone. China clay. Glass. Paper. Plaster. Cement. Phosphate fertilizers.

8.1 INTRODUCTION

This chapter is restricted to a discussion of four materials which are either used in the home, e.g. paper, or constitute parts of the home, e.g. glass, plaster and cement, and the phosphate fertilizers which are vitally important in food production. The manufacture of paper from wood pulp is dependent upon the availability of water, sulfuric acid, sodium sulfite (Na_2SO_3), sodium hydroxide, and china clay or kaolin. The glass industry also makes use of china clay, but has quartz sand and limestone as major ingredients of its products. The cement and plaster industries are dependent upon limestone as a basic substance. Before the composition and production of paper, glass, plaster and cement are described in detail, the material, sand, and the basic substances, limestone and china clay, are described.

8.2 SAND

The geological definition of sand depends upon the particle size distribution of the material. Sand grains are produced by the chemical, biological and physical weathering of rocks. As particles of broken rock are produced by these erosion processes, the particle size is reduced as the material is washed down rivers and thence to the sea. The grains become more rounded as the process continues and only those particles within the size range 0.06 - 2 millimetres diameter are called sand. Larger particles are known as gravel (2-60 mm), pebbles (60-200 mm) and boulders (greater than 200 mm). Smaller particles are called silt (0.002 - 0.02 mm) and clay (less than 0.002 mm). The sand most commonly used consists mainly of quartz which is silicon dioxide, SiO_2. Quartz is a very hard mineral which is glass-like in appearance but can be white or coloured depending upon the presence of impurities. Minor components of sands are the minerals, mica, feldspar and calcite, and the rocks, slate, basalt and limestone. Sand deposits are found in various locations where the material has collected at the foot of hill slopes, some of which have been buried and converted to sandstone by the high pressure and various 'cementing' deposits from aqueous seepage which include iron(III) oxide and calcium carbonate. The sand used industrially is obtained either by open quarrying or by underground mining. In the case of sandstone, the sedimentary rock has to be crushed and ground down to the suitable size range. For building purposes, the sand is not subjected to any further treatment. The sand used for glass manufacture must be almost pure quartz or silica sand which is found in particular deposits, usually those produced by glacial movement. Even so, the sand from these deposits must be purified by washing with sulfuric acid so that it contains at least 98.5% silica before being used for glass manufacture. The product has a restricted size range between 125 and 500 μm in diameter, which is essential to ensure even melting in the glass making process. The formula of silicon dioxide, SiO_2, is very simple but represents a giant structure in which each silicon atom is bonded to four tetrahedrally arranged oxygen atoms and an oxygen atom bridges between every closest pair of silicon atoms. The structure of one common form of silica, known as β-cristobalite, has a close relationship with that of diamond, the structure of which is shown in Fig. 11.5. The silicon atoms have the same arrangement as do the carbon atoms in diamond with every closest pair participating in a silicon-oxygen-silicon bridge, the Si-O-Si bond angle being 144°.

8.3 LIMESTONE

Calcium carbonate, $CaCO_3$, exists naturally as marble, limestone, calcite, aragonite and chalk. The various forms differ in their origins and crystalline make-up. Calcite and aragonite are names given to pure crystals which differ in the manner in which the calcium ions and carbonate ions are arranged, the former being very slightly more stable. Although all calcium carbonate minerals are fairly soft as rocks go, aragonite is somewhat harder and denser than calcite. It is the preferred form at high pressures and temperatures. Limestone is the massive form of calcite which has crystallized out from aqueous solutions originally containing calcium ions and hydrogencarbonate ions. Forms of limestone are found as stalactites that hang down, and stalagmites that point upwards, in caves through which water seeps. The shells of marine organisms consist largely of the aragonite form of calcium carbonate and form the basis of reefs and chalk deposits. The metamorphosis, which consists of the processes of dissolution and recrystallization, of limestone under conditions of high pressure and temperature leads to the formation of marble. In its purest form marble is white, e.g. as is the deposit at Carrara in Italy, used by Michaelangelo and others for sculpture, but silicate impurities cause it to have attractive variegated coloured veins. The various deposits of limestone in the Earth amount to around 20,000 million million tonnes in keeping with the position of calcium as the fifth most abundant element; 44600 grams per tonne of the Earth's crust.

8.4 CHINA CLAY (KAOLIN)

China clay or kaolin is a weathering product of feldspar. Orthoclase or potassium feldspar has the formula, $K_2O.Al_2O_3.6SiO_2$, expressed in terms of oxide composition. The potassium oxide content is washed away as soluble potassium hydroxide in the weathering process leaving the rock containing the mineral kaolinite, $Al_2(OH)_4Si_2O_5$, which is called china clay or kaolin. The name, kaolin, is derived from the Chinese for high hill, the material having been first discovered in the weathering products of a mountain in China. The clay is usually quite pure and practically white which makes it attractive for the manufacture of fine china and porcelain and for incorporation into paper.

The world production of kaolin in 1990 was 25 million tonnes worth $2.5 billion. The distribution of world production is shown in Fig. 8.1.

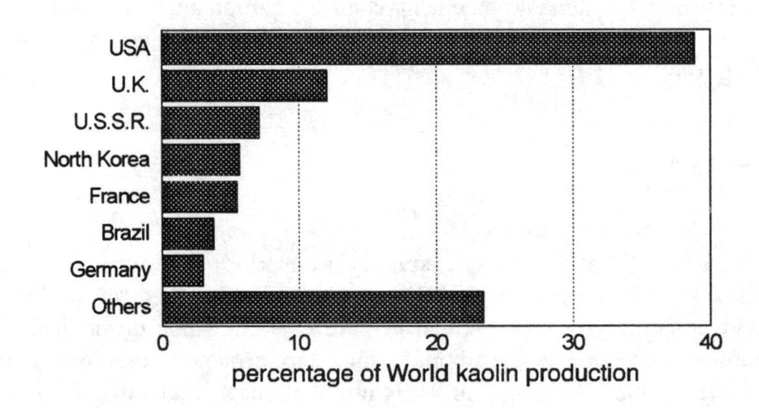

Fig. 8.1 The distribution of World kaolin production in 1990

In spite of being discovered in China, the annual production of kaolin in that country is less than 0.2% of the world supply and there are only reserves of around 200 million tonnes there, which is not much more than that produced in the world in one year at present rates. The world reserves are adequate for around 7800 years! The major use (43%) of kaolin is in the filling and coating of paper. Other uses are in the production of refractory materials, e.g. furnace linings, ceramics, glass, paint and rubber.

8.5 GLASS

Glass was discovered, together with the manufacture of cooking pots by the baking of clay, shortly after the discovery of fire. Glass is a normally transparent, hard brittle substance which is mainly composed of silica. Other oxides are incorporated in various proportions to produce glasses with differing properties. Sodium oxide, Na_2O, is introduced by fusing silica sand with soda ash, Na_2CO_3, the carbonate decomposing at the high temperatures used, to give the oxide. In a similar manner any calcium oxide (CaO) content is added as limestone, $CaCO_3$. Alumina, Al_2O_3, is a constituent of some sands, but any additional material is derived from a basic slag from the steel industry called calumite which contains 15% of alumina. It does contain small quantities of iron and can be of limited use. Any additional alumina is derived by adding a natural mineral silicate called nepheline syenite. The addition of boric oxide, B_2O_3, produces borosilicate glasses. Pure silica melts at 1713°C, some fused silica glass being used for photochemical reaction vessels since it transmits ultra-violet radiation. A translucent variety of fused silica known as Vitreosil™ is used in chemical industrial plant and for some domestic cooking utensils. It is resistant to both thermal shock and to acidic attack. Transparent silica glass has limited use as a window material in the construction of summer houses in which the ultra-violet irradiation of people may be carried out without their having to endure the ravages of the wind. For normal purposes of keeping the wind and rain out of buildings and allowing the transmission of visible light the much cheaper soda glass is used.

The addition of soda ash, limestone and calumite, to the quartz sand reduces the melting point of the mixture to around 1500°C. Soda glass used for windows has a composition of 73% silica, 13.5% sodium oxide, 11% calcium oxide and 2.5% alumina. Once the carbonates have decomposed to give oxides, the molten glass can be worked at temperatures as low as 600°C. A typical container glass would have the composition 73% silica, 15% sodium oxide, 10% calcium oxide and 2% alumina. Lead crystal glass has the composition 55% silica, 32% lead oxide (PbO) and 13% potassium oxide. The soda glasses are subject to breakage when heated quickly, their coefficient of thermal expansion being around 8.5 micrometres per centimetre per degree Celsius. The comparable figure for copper metal is 17 μm per cm per degree, but borosilicate glass (Pyrex™) containing 80.5% silica, 12% boric oxide, 8.2% of alumina, 4.5% sodium oxide and 0.8% zinc oxide is more able to withstand rapid heating with a value of only 3 μm per cm per degree Celsius. Borosilicate glasses are used for the manufacture of chemical apparatus, domestic utensils and television tubes. If potassium carbonate is used instead of soda ash, the product is a potassium glass, such material being used for crystal glasses as it is harder and more brilliant than soda glass. Lead crystal is even more brilliant. The brilliance of a glass is dependent upon the value of its refractive index which, in turn, is dependent upon the composition of the glass. Refractive indices of glasses vary between values of 1.5 to 1.9, the figure for diamond, an element valued for its brilliance, being 2.42.

Although the individual components of glass are compounds which have highly ordered crystal structures, the fused mixture is unable to become as ordered as a crystal. As it cools down, the glassy state is formed. As the mixture cools the viscosity increases to an extent which prevents the formation of the long-range order typical of the crystalline state. There is some short-range order in a glass but it is that which is more typical of the liquid state. A glass may be regarded as a super-cooled liquid, very old glass eventually becoming devitrified as more long-range order is established. In these circumstances the glass becomes crystalline and extremely brittle. The lack of long-range order causes glass to have a softening point rather than a melting point so characteristic of crystalline substances. As glass is heated, it becomes less viscous and above the softening point it may be worked and shaped.

About 7% of food and beverage packaging consists of glass containers, a figure which has steadily been decreasing for some time as the result of competition from other packaging materials with plastic being the most serious. The production of container glass in Europe in 1991 amounted to 15.7 million tonnes, six million tonnes of which represented recycled material. The major European manufacturers of container glass, together with their percentages of production in 1991 are shown in Fig. 8.2.

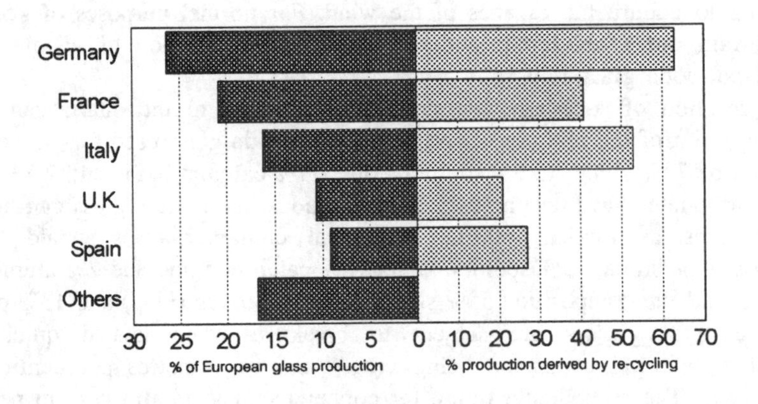

Fig. 8.2 The distribution of glass manufacture in Europe and the percentages of the production which were derived from recycled glass (1991 data)

Also included in the diagram are the percentages of the production which were derived from recycled glass. The competition from other container materials and the increasing recycling rate have combined to make the use of primary materials level off. Containers made from plastics amounted to 38% of the European total and represent a considerable use of non-renewable chemical feed stocks (oil). In 1979 the figure was 22%, indicating the high growth of the use of plastics by the packaging industry.

8.6 PAPER

Paper is reputed to have been invented in China around the second century AD. The spread of the technology throughout the rest of the world was very slow, the Moors in Spain having the knowledge in the twelfth century. The news spread to France (1189), Italy (1260), Germany (1389), Switzerland (1400), Belgium (1407) and Holland (1428), reaching Great Britain in 1490. Afterwards it spread to Sweden (1532), Denmark (1540) and to Russia and the United States in 1690.

The principal component of paper is cellulose, which is the fibrous material responsible for the strength of plant stems and trees. It is the main component of cotton and linen fibres out of which paper used to be made until the 19th century. Even today, waste cotton is incorporated in some papers. The source of cellulose for the modern paper industry is the wood pulp derived from the fast growing softwood forests of Scandinavia and North America. In recent years, wood plantations in Argentina, Brazil and Chile have begun to contribute to the world supply of wood pulp. In these countries with a warmer climate, trees grow more quickly and land is cheaper. In general, trees grow for 100 days per year in Scandinavia and every day of the year in Brazil. Eucalyptus trees are in a suitable state for felling after five years growth. This is half the time for a similar growth of spruce in Scandinavia to mature. In Chile, use is made of the Radiata pine which matures in fifteen years as opposed to 40 years in Sweden and 60 years in Canada. Some 15% of world pulp supplies originate in Latin America, a figure which is expected to increase. Significant developments in pulp manufacture are occurring in Thailand, Malaysia and Indonesia. It has been estimated that 5% of Indonesia's forests, if turned into properly managed plantations, could account for one third of the world's pulp supplies in future.

Cellulose constitutes up to 49% of wood and is surrounded by a material called lignin which accounts for 30% of the mass. The lignin adds strength and rigidity to the cellulose fibres. In the process of pulping, lignin is removed from the wood so that the cellulose fibres may be used for papermaking. The main process used for wood pulp manufacture is the Kraft™ (the word kraft means strength in German and refers to the cellulose fibres produced by the process) or 'sulfate' method in which the softwood logs are converted to chips and then digested with an alkaline sodium sulfide solution. In the process the sulfide ions (S^{2-}) are converted to sulfite ions (SO_3^{2-}) and the lignin is dissolved. The cellulose fibres are then removed by filtration leaving the solution to be regenerated. The solution is evaporated to dryness and the solid product is roasted after the addition of sodium sulfite. The roasting causes the residual wood to be converted into carbon dioxide, which reacts with the hydroxide ions to give carbonate ions. Some of the carbon from the residual wood reduces the added sodium sulfate and the residual sodium sulfite to sodium sulfide. Addition of lime (CaO) removes the carbonate ion as calcium carbonate, leaving a solution of sodium hydroxide and sodium sulfide for the treatment of the next batch of wood. The separated calcium carbonate is heated to form lime. All the chemicals are recycled as far as is possible with the addition of sodium sulfate at each stage to make up for the inevitable losses. The use of sodium sulfate gives rise to the name of the overall process although that substance is the precursor of the sodium sulfide which actually attacks the lignin. There is also a lesser used acidic sulfite process for the manufacture of wood pulp. Wood pulp is usually manufactured in the countries which have the forests and is then exported world-wide to those countries which manufacture paper.

In the paper mill, wood pulp bales are re-pulped in water so that the resulting suspension of bundles of fibres can be beaten. Beating is a process in which the fibres of cellulose are separated and treated for conversion into papers as different as absorbent blotting paper and grease proof paper. Individual fibres are hydrated, which allows some natural sizing to occur, fluffed, which enables the fibres to bond together more strongly in the papermaking stage, and shortened. When the appropriate amount of beating has taken place the fibre slurry is allowed to enter the papermaking machine. The slurry is spread out over a 9 metre wide continuous belt of copper wire gauze, known as the 'wire', moving at up to 1000 metres per minute. The majority of the water is removed from the fine layer of what is to become paper by a combination of gravity and suction. At the end of the horizontal section of the wire the continuous sheet of paper is fed through heavy press rollers to remove more water and continues along the machine through a series of dryer rollers which are heated by steam and which dry the paper so that it can be collected as a roll in the final stage. At the beating stage sizing agents may be added to the slurry as well as filling materials. The sizing agents are added to control the resistance to absorption of water so that printing inks do not spread as they do on newsprint. The fillers are added to improve the opacity and whiteness of the final product. The most used filler is china clay (kaolin), although considerable quantities of talc, calcium carbonate and titanium dioxide are important alternative substances. Paper which is used for glossy magazines and food labelling is given further treatment before it leaves the paper mill. The gloss is applied to the paper in the form of a slurry of finely divided filler substances, the coated paper being subjected to super-calendering by passing through a stack of rollers which are alternately, heated highly polished steel and resilient fabric, e.g. asbestos. The frictional effect of passing through the stack of rollers is to produce a very high polished finish to the paper which improves its appearance and its receptivity to printing ink.

Recycled waste paper contributes up to 25% of the paper and cardboard manufactured by some countries. A small fraction of waste paper can be admixed with new pulp without affecting the properties and quality of the finished products. Paper made from recycled paper is of poor quality because the extra beating the feed experiences shortens the fibre length too much to produce a properly cohesive product. The colour of recycled paper is far from perfect and its receptivity to printers ink is low. Paper waste is more appropriately used for the manufacture of cruder materials such as card and cardboard. The forests which produce trees for wood pulp manufacture are more than sustainable as more trees have been planted than are used for many years now. The Swedish stock of suitable timber is rising at the rate of twelve million tonnes per year at present and that rate of increase is expected to continue until at least the year 2060. The world production of wood pulp in 1991 was around 35 million tonnes, a figure which is in excess of demand by 5 million tonnes due to the current world recession. The relative amounts of the various paper and board produced are shown in Fig. 8.3.

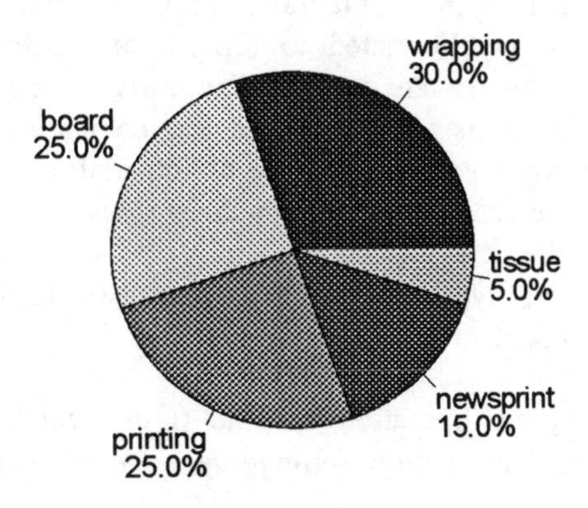

Fig. 8.3 The relative amounts of the various paper and board produced

8.7 PLASTER

Plasters have as their basis the material known as 'plaster of Paris' which is manufactured from the natural mineral gypsum, $CaSO_4.2H_2O$, or calcium sulfate dihydrate. Gypsum exists as a sedimentary rock which is fairly pure. When finely powdered gypsum is heated to between 120-160°C it loses some of its water of crystallization to give the hemi-hydrate, $CaSO_4.{}^1/_2H_2O$, which is the formula of plaster of Paris. At temperatures above 200°C all the water is removed to give the anhydrous calcium sulfate which is called anhydrite. Plaster of Paris sets very quickly when mixed with water, the reaction being the re-hydration of the hemi-hydrate to give gypsum. It is used in the plaster bandages which are used to set fractured bones. Anhydrite reacts more slowly with water in producing gypsum and is much easier to use than plaster of Paris. In practice there is a range of plasters available which are mixtures of plaster of Paris, anhydrite and slaked lime, $Ca(OH)_2$. The addition of slaked lime slows down the reaction with water and over the course of time is converted into calcium carbonate by reaction with atmospheric carbon dioxide. This causes a long term hardening and strengthening of the final solid plaster.

8.8 CEMENT

Portland cement, so-called because of its resemblance to Portland stone, which is limestone, when set, is a complex mixture of substances manufactured by heating limestone, sand and clay, which is an aluminosilicate, in a kiln at temperatures between 1450-1600°C. The resulting solid clinker consists of a mixture of calcium silicates, calcium aluminate and calcium aluminate ferrate. The tricalcium silicate, $3CaO.SiO_2$ (abbreviated to C_3S), known as alite (*a-lite*) and dicalcium silicate, $2CaO.SiO_2$ (abbreviated to C_2S), known as belite (*bee-lite*) are the cementing agents. Calcium aluminate, $3CaO.Al_2O_3$ (abbreviated to C_3A), and the calcium aluminate ferrate, $4CaO.Al_2O_3.Fe_2O_3$ (abbreviated to C_4AF), are together called celite (*cee-lite*) and are fluxes. The fluxes are produced initially in the cement kiln and form a liquid medium for the proper formation of the cementing agents. The chemical reactions which occur when water is added to cement are very complex but are understood in outline. The reaction occurs in three stages.

(i) The first stage is the hydration of the two fluxes C_3A and C_4AF which is relatively rapid, taking about one hour.

(ii) The second stage is the hydration of alite (C_3S) which produces belite (C_2S) and calcium hydroxide. This is the setting stage which takes up to twenty four hours.

(iii) The third very slow stage consists of two reactions, one of which is the hydration of the belite (that initially present and that produced in the hydration of the alite), the other being a reaction of the hydrated C_3A flux with the calcium hydroxide. Both the third stage reactions cause hardening of the cement which can take at least one year to complete. If there is insufficient free calcium hydroxide to allow the latter reaction to occur the hydrated C_3A flux removes calcium oxide from the belite, causing it to be degraded to compounds such as C_3S_2 and CS which hydrate rapidly and interfere with the desired slow third stage hardening reaction of the belite. To avoid this, a proportion of gypsum, $CaSO_4.2H_2O$, is added to the powdered cement. The gypsum reacts preferentially with the hydrated C_3A flux and so prevents it from decalcifying the belite.

Cement is used normally with an admixture of either sand to make mortar or with sand and gravel or building rubble to make concrete. The resulting mixtures set to give conglomerates, the inert materials being bound together by the solidified cement. Gravels which may be used for concrete mixes vary in particle size between 5-60 millimetres in diameter. The best sand for use in mortars is that known as sharp sand, the individual particles being less rounded than those of normal sand. The sand in mortars used for bricklaying should pass through sieves with a mesh size of one millimetre. That used in concrete mixes should pass through sieves with a mesh size of 5 millimetres. Good sand may be identified by its sharp feel and when washed should not produce any turbidity in the water.

8.8 PHOSPHATE FERTILIZERS

Phosphate fertilizers are manufactured from phosphate rock which contains phosphorus in the form of fluoroapatite, $Ca_5(PO_4)_3F$. The world production of phosphate rock in 1990 amounted to 154 million tonnes containing the equivalent of 47.5 million tonnes of phosphorus pentoxide, P_2O_5, and was worth \$5.6 billion. The distribution of world production of phosphate rock in 1990 is shown in Fig. 8.4.

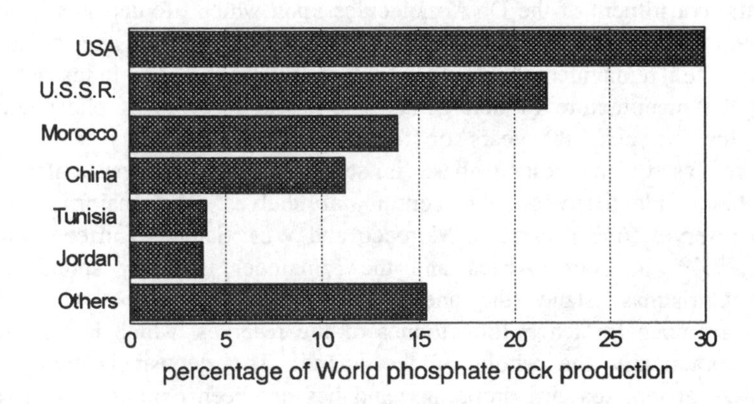

Fig. 8.4 The distribution of World production of phosphate rock in 1990

Dissolution of phosphate minerals by rain water causes around two million tonnes of the material to enter the seas annually. Man's activities roughly double this quantity and in the U.K., for example, detergents account for 100,000 tonnes, human excreta for 75,000 tonnes and 25,000 tonnes by various industrial processes being released into the country's sewers.

The fluoroapatite in phosphate rock is converted to a useful form by treatment with sulfuric acid:

$$2Ca_5(PO_4)_3F + 7H_2SO_4 + H_2O \rightarrow 7CaSO_4 + 3Ca(H_2PO_4)_2.H_2O + 2HF$$

The calcium dihydrogen phosphate monohydrate is reasonably soluble in water and the two calcium salts are sold together as 'superphosphate' fertilizer, the calcium sulfate being a useless insoluble component. This may be avoided by the conversion of fluoroapatite into 'triple superphosphate' by treatment with phosphoric acid:

$$Ca_5(PO_4)_3F + 7H_3PO_4 + H_2O \rightarrow 5Ca(H_2PO_4)_2.H_2O + HF$$

the solid fertilizer containing around three times as much soluble phosphate then ordinary superphosphate, justifying its name. The phosphoric acid used in the process has to be made from fluoroapatite by treating it with sufficient sulfuric acid to convert all the calcium content into gypsum:

$$Ca_5(PO_4)_3F + 5H_2SO_4 + 10H_2O \rightarrow 5CaSO_4.2H_2O + 3H_3PO_4 + HF$$

The insoluble gypsum is filtered off, the fluorine is precipitated as insoluble sodium fluorosilicate, Na_2SiF_6, by the addition of sodium silicate and the resulting solution is concentrated by boiling off the water.

The importance of phosphorus in the diet and the role of the element as a fundamental constituent of the DNA molecules upon which life depends is described in Chapter 9. About 95% of phosphate production is used in the manufacture of fertilizers, the remainder being used for a variety of industrial processes including the manufacture of detergents. The world reserves of phosphate rock are adequate for the next 80 years or for more than 200 years if less accessible deposits are used. The reserve base is steadily increasing, particularly with the deposits being identified on the continental shelves. The majority of currently identified reserves (63%) reside in Morocco and West Sahara. Thirteen percent lie in the USA, 7.4% in South Africa and the remainder is thinly spread over many countries. Christmas Island (the one in the Indian Ocean and which is part of Australia) accounts for ten million tonnes of the reserves, which is only 0.03%, but represents practically the whole of the island! The deposit is thought to have originated as guano, sea-bird droppings, and has not been exploited since 1987.

9

Food and life

Energy content of food. Energy requirements for various activities. Nutritional data. Composition of the body. Essential elements. Proteins, amino acids and enzymes. Carbohydrates and sugars. Fats, fatty acids and glycerol. Vitamins. Hormones. Cholesterol. Trace elements. Food metabolism. DNA, protein synthesis, life and genetics.

9.1 FOOD AND ENERGY

To a chemist, food is a variable mixture of carbohydrates, proteins, fats and essential trace elements and compounds, including hormones and vitamins. Section 9.3 is devoted to chemical descriptions of the components of food. These components contain the building materials which are necessary for the construction and maintenance of the cells from which humans are constructed and the compounds which allow them to operate efficiently. People are constructed from, and are maintained by, the components of the materials which they eat and drink. The contributions of the main constituents of prepared and canned foods; carbohydrates, proteins and fats, are usually indicated on the label of the packaging material in terms of their calorie equivalents. There is a possibility of some confusion about the units of energy quoted on foods. One calorie, with a small *c*, is the energy required to cause the temperature of one gram of water to increase by one degree Celsius. The Calorie, with a large *C*, is that commonly used on food labels, although it sometimes written with a small *c*, and is really the kilocalorie, usually written as the abbreviation, 'kcal'. The proper scientific unit is the kilojoule, abbreviated as 'kJ', the kilocalorie being equal to 4.184 kilojoules (kJ) of equivalent energy. These relationships are summarized as:

$$1 \text{ Calorie} = 1000 \text{ calories} = 1 \text{ kilocalorie} = 4.184 \text{ kilojoules}$$

A normal healthy male with a weight of 80 kilograms (176 lb) and a height of 176 centimetres (70 inches) requires 60 grams of protein per day for body maintenance. A similarly average female (63 kg or 138 lb weight, 163 cm or 64 inches height) requires 50 grams of protein per day. Actual diets vary widely, most people in the western world overeating considerably. The average persons just described, leading an average existence, need intakes of energy no greater than 2500 Calories (2500 kcal or 10460 kJ) and 1650 Calories (1650 kcal or 6900 kJ) respectively. A measurement which combines the height and weight of a person so that it relates to the person's health status is the **body mass index**. This is obtained by dividing the person's weight in kilograms by the square of their height in metres. The average male described above would have a body mass index of:

$$\frac{80}{1.76 \times 1.76} = 25.8$$

The body mass index of the average female is:

$$\frac{63}{1.63 \times 1.63} = 23.7$$

The acceptable range of values of the body mass index, independently of gender, is between 20 and 25. The value for the average male is outside the acceptable range and emphasizes the point made above that many people, mainly males, eat too much for the exercise they take. The average female has a body mass index in the middle of the acceptable range.

There are many books which list the calorific values of virtually all possible foodstuffs, and a great deal of dietary information comes with packaged foods. If the daily intake of a person is greater than the body needs, the person puts on weight. If the intake is less than the body needs, the person's weight decreases. People who wish to slim have to bear such numbers in mind and exercise will-power in order to succeed. No amount of proprietary slimming aids will do any good unless those conditions are adhered to. If the conditions are met, no proprietary aids should be required.

It is not intended that this book should compete with books about nutrition, and the many slimming manuals that are available. However, because food and its fates are important from the chemical point of view, some consideration is given to the calorific values of a limited range of foods and drinks. Foods and drinks have calculable energy equivalents; the energies which they would produce if completely oxidized. Some of the energy derived from food is used in the maintenance and operation of the body and some is released as heat. Any excess of energy intake over that used is stored in the form of fat. The intake values mentioned above are those which would allow maintenance and operation of the body at the expense of a small amount of the energy stored as fat and would allow gentle slimming to occur. The make-up of the minimum levels of food and drink intake which would allow slimming is almost infinitely variable but some data are given in the following tables with appropriate discussion.

People who engage in sporting activities and professional athletes require energy intakes up to 50% greater than the amounts quoted above, except for Sumo wrestlers who are too busy eating everything in sight to worry about the definitions of energy units and intake statistics. The average hourly requirements of energy for men and women taking part in various activities are given in Table 9.1.

Table 9.1 - Average hourly energy requirements in kilojoules for males and females taking part in some activities

activity	males	females
sitting	470	350
walking	940	700
running	2475	1860
cycling	1070	800
gardening	1540	1150
tennis	1875	1400
skiing/squash	2500	1900
swimming hard	2680	2000
cycling hard	2760	2100
rowing hard	3350	2500
running hard	3770	2900

The major basic energy-rich components of food and drink may be classified as protein, carbohydrate and fat. The carbohydrate content of food may be divided into that which is available to the body and that which is unavailable. Unavailable carbohydrate is mainly found in fibre which cannot be metabolized, but is very important for the regular and comfortable action of the bowels. Whole books have been written about the beneficial effects of eating fibre, but no further treatment is given in this book. The carbohydrates that are discussed in this chapter are those that contain available energy. Protein is needed for body cell building and maintenance, carbohydrate is needed for operational energy and fat is needed for membrane and tissue construction and as a long-term energy store. Excesses of fat are unnecessary and lead to an overloading of the body's operational capacity.

Energy equivalents of the three basic food components, together with those for the commonly used alcoholic drinks, are given in Table 9.2. Alcoholic drinks contain no fat and negligible amounts of carbohydrate and protein, deriving their energy content exclusively from their alcohol content, the alcohol being ethyl alcohol or ethanol, C_2H_5OH. The energy content of any food or beverage is estimated in terms of its heat of combustion, i.e. the amount of heat produced by burning it, expressed as kJ per 100 grams (or 100 mL for liquids). Pure ethanol has a heat of combustion of 1371 kJ per mole (one mole of ethanol is 46 grams = 58 mL, contained by 145 mL of a 40% spirit drink), an energy equivalent of 2363 kJ for 100 mL. Any alcohol intake is essentially very available energy and prevents the use of an equivalent amount of food. The food which is surplus to the body's requirements is then diverted into fat production.

Table 9.2 - Energy equivalents of basic food components and alcoholic drinks

substance	energy equivalent of 100 grams (100 mL for liquids) in kJ
carbohydrate	1570
protein	1690
fat	3730
beer	125
wine	280 (dry) - 390 (sweet)
whisky (40%)	945

The data given in Table 9.2 are relevant to those given for individual foods in subsequent tables, the data for the alcoholic drinks should be considered when calculations of the intake of energy per day or per meal is carried out. The values for spirits other than whisky have the same energy content as whisky, but differ significantly in taste. A balanced diet of food for the average male person is considered to consist of 80 grams of protein, 350 grams of carbohydrate and 100 grams of fat, although considerable variations from these figures will be found from various sources. The appropriate figures for females are 53 grams of protein, 230 grams of carbohydrate and 66 grams of fat.

The following sections include a brief treatment of the problems of nourishment and are not intended to supplant the excellent books about diet which are available. They are included as a general guide to the amount of energy taken in by humans on a daily basis and to demonstrate how easily the essential intake may be exceeded.

Some substances which may form a person's breakfast, their energy equivalents and their contents of carbohydrate (C), protein (P) and fat (F) are given in Table 9.3.

The contributions to the total weights of the materials in Table 9.3 do not add up to 100 grams, mainly because of the omission of the weights of water. There are also minor constituents which may be very important but they represent a very small part of the weight and are dealt with below. This comment applies to all the tables of data in this chapter. If bread is used, the butter or margarine spread upon it has to be considered. Butter and margarine have the energy equivalent of fat as indicated in Table 9.2. They differ in the distribution of the fat content between saturated, mono-unsaturated and poly-unsaturated molecules. Butter contains cholesterol, but margarine has none. The differences between butter, hard and soft margarines are given in Table 9.4.

Table 9.3 - Nutritional data for some breakfast foods (100 g of solids or 100 mL of liquids)

food	energy/kJ	C/g	P/g	F/g
orange juice	185	11	0	0
cornflakes	1645	86	7	0
oat & bran flakes	1270	61	12	2
milk (full cream)	850	24	4	11
Weetabix™	1448	68	12	3
bacon	775	2	24	9
eggs	670	2	12	12
kippers	205	0	26	11
bread (wholemeal)	1020	45	10	4
jam	1150	70	0	0

Table 9.4 - The fat and cholesterol contents of 100 gram quantities of butter and margarines

substance	butter/g	hard margarine/g	soft margarine/g
saturated fat	51	16	14
mono-unsaturated fat	23	36	29
poly-unsaturated fat	3	25	35
cholesterol	0.22	0	0

Lunch is restricted, in this treatment of the subject, to a sandwich of cheddar cheese, corned beef or sardines, the first of which should be avoided by slimmers as may be deduced from the data given in Table 9.5.

Table 9.5 - Nutritional data for 100 gram quantities of some sandwich ingredients

food	energy/kJ	C/g	P/g	F/g
cheddar cheese	1720	0	25	32
corned beef	910	0	26	12
sardines in oil (drained)	877	0	24	13

The data for some foods which may be used at dinner time are given in Table 9.6.

Table 9.6 - Nutritional data for 100 gram quantities of some starters, main course ingredients and desserts

food	energy/kJ	C/g	P/g	F/g
peanuts (roast)	2510	7	25	53
potato crisps	2140	40	6	37
avocado pear	740	7	2	17
tomato soup	150	7	1	1
clam chowder	140	5	2	1
beef	870	0	31	8
lamb	1010	0	26	15
pork	1320	0	28	22
chicken	1090	9	25	13
turkey	720	0	29	5
duck	840	0	24	11
salmon	690	0	25	6
prawns	490	1	25	1
potatoes	460	25	2	0
pasta	130	26	5	1
peas	170	7	3	0
carrots	190	10	1	0
beans (butter)	430	18	7	1
onions	120	6	1	0
apples	240	15	0	0
bananas	390	24	1	1
apple pie	1640	43	7	22
fruit cake	1590	58	5	17
cream (thick)	1440	3	2	37
oatcakes	1845	63	10	18
chocolate	2190	65	5	29

Roasted peanuts and potato crisps are included as they may feature as a pre-dinner sources of energy and as a dire warning to anyone with aspirations of losing weight. Peanuts represent a very good source of all three major food components. An intake of 420 grams would supply the complete daily requirement for a person needing 2500 Calories (10460 kilojoules) and would be an adequate supply of the necessary protein, although the carbohydrate content is much lower than that recommended for a balanced diet. Such a quantity of peanuts would cost around £2.20 ($3.40). No cutlery or crockery would be required and the rate of intake of nuts could be varied from between 1-3 eating sessions (it would be a misnomer to call them meals) to an almost continuous rate of three small nuts every five minutes over a twelve hour period. Persons living (?) on a diet of peanuts would need to supplement their intake with one tablet per day of multi-vitamins with minerals (costing 7p or ten cents per tablet) and drink a reasonable amount of water, particularly if the peanuts were of the salted variety. Or, how about 566 grams of oatcakes per day? Again, the vitamin/mineral tablet would be necessary. Another possibly attractive diet would be three meals, each consisting of a bowl of cornflakes or three Weetabix™ biscuits and a tin of sardines. This would have sufficient vitamin content to make the tablets unnecessary.

The avocado, not to be confused with Avogadro, pear is often used as a starter for a meal but its high fat content means that slimmers should opt for the soup or go straight to the main course. For the protein intake, slimmers would be advised to eat either beef, poultry or fish. Carbohydrate is necessary for immediate conversion into usable energy, potatoes and pasta being good sources. The other vegetables are eaten mainly for their mineral and vitamin contents rather than as a source of energy. It would be possible to derive as much energy as is contained by 100 grams of beef by eating around 500 grams of peas or 720 grams of onions. Slimmers should consume fruit salad (no cream) for their dessert, apple pie or fruit cake being far too energy-rich for them. Persons taking part in activities which require the consumption of large amounts of energy should derive most of the extra energy from sources high in carbohydrate, e.g. pasta, such materials being most easily converted into usable energy.

9.2 ELEMENTS OF THE HUMAN BODY

The essential elements, from which the human body is composed, may be categorized in terms of their percentage contributions to the body's mass. The four major essential elements, together with their percentage contributions to the body's mass, are listed in Table 9.7.

Table 9.7 - Percentage contributions to body mass by the four major essential elements

element	percentage
oxygen	64.8
carbon	18
hydrogen	10
nitrogen	3.1

They account for 95.9% of body weight and are the elements which make up the 'organic' components of the body. The second group of essential elements make up the other 4.1% of body weight are listed in Table 9.8 and may be regarded as the minor elements in terms of their contributions to the body's weight.

Table 9.8 - Percentage contributions to body mass by the minor essential elements

element	percentage
calcium	1.97
phosphorus	1.08
potassium	0.37
sulfur	0.26
chlorine	0.17
sodium	0.11
magnesium	0.04
iron	0.01

About 90% of the calcium in the body is located in the bones and teeth, the remainder is contained by the blood. It is concerned with the blood clotting mechanism, with muscle contraction and heart function, and is necessary for the proper utilization of iron. Most of the calcium in the diet is supplied by dairy products (milk, cheese and yoghurt), vegetarians having to obtain an adequate supply by taking calcium tablets. The recommended daily intake of calcium is between 800-1200 mg in adults. An excessive intake of soft drinks containing phosphoric acid (colas) can decrease calcium absorption by the formation of insoluble calcium phosphate. However, it is most unlikely that cola would be drunk at the same time as the breakfast cereal and milk or that milk and cola would be drunk together at any other time.

As with calcium, 90% of phosphorus in the body is located in the bones and teeth. This is to be expected since the main inorganic constituent of bone is calcium phosphate. The remainder of the phosphate content is distributed between various enzymes, phospholipids which are a constituent of cell membranes, and as part of the DNA (deoxyribonucleic acid) and RNA (ribonucleic acid) molecules which govern protein production, thus ensuring the maintenance of cells and reproduction. Another pair of molecules vital to body function are adenosine triphosphate (ATP) and adenosine diphosphate (ADP) which lose and gain a phosphate group (P) respectively, in ATP hydrolysis and formation:

$$\text{ATP} + H_2O \rightleftharpoons \text{ADP} + \text{P} + \text{energy}$$

The interchange of the phosphate group is extremely facile and leads to the release of energy when ATP is transformed into ADP, a process which facilitates almost all of the metabolic functions of the body. ATP is normally complexed with calcium ions. The recommended daily intake of phosphorus is 800-1200 mg but its prevalence in foods ensures that no supplementation is required. No cases of deficiency have been reported.

Potassium, present in the body as the aqueous ion, $K^+(aq)$, is vital for the maintenance of cell function. It is concerned with the mechanism of energy release in cells, the manufacture of proteins, it regulates fluid balance and the heart beat. It regulates the pH value of the blood by exchanging with aquated protons when necessary in an ion-exchange process. It takes part in the transmission of nerve impulses and in muscle contraction. This highly important element is present in a wide variety of foods so that deficiency is rare. Deficiency can occur in illnesses which cause prolonged diarrhea but this can be rectified by taking potassium chloride tablets.

Sulfur is present in every cell as the component of the amino-acid units of proteins, cysteine and methionine. It is a constituent of enzymes, its functions being to maintain the hair, nails and skin and to participate in the maintenance of body metabolism. Protein foods (meat, fish, nuts, beans) together with cabbage, broccoli, cauliflower and Brussels sprouts, are all good sources of sulfur.

Chlorine, as the aqueous chloride ion, $Cl^-(aq)$, is a very important constituent of cells and body fluids. It is crucial to the working of the stomach where it is present as hydrochloric acid. Dietary deficiency is almost unknown, but cases of severe dehydration and heat illness are known and are treated with a suitable solution of sodium chloride or with salt tablets. Chlorine is found in a variety of foods including green vegetables, tomatoes, asparagus, celery, parsnips, carrots, onions and pineapples. It is added to most foods as salt, sodium chloride (NaCl), in the cooking or just before they are eaten.

Sodium, as the aqueous ion, $Na^+(aq)$, is concerned with several important functions including the maintenance of normal fluid balance, together with potassium - the sodium:potassium ratio being critical, the transmission of nerve impulses, muscle contraction and the regulation of pH of body fluids. It is known that sodium raises the blood pressure of those people who suffer from hypertension, i.e. higher than normal blood pressure, and should be restricted in the diet of people with cardiovascular disease. As with potassium and chlorine, deficiency can occur in cases of severe dehydration but is easily countered. The addition of salt to food improves the taste and there is no possibility of deficiencies occurring normally. On the contrary, it has been estimated that the normal diet contains 2-3 times more than is actually required by the body for its normal functions. Blood normally contains sodium, potassium and chloride ion concentrations of 136-142 millimolar (mM), 3.8-5 mM and 95-103 mM respectively, with the charge balance being maintained by hydrogencarbonate ion (bicarbonate, HCO_3^-) and small concentrations of magnesium, calcium and phosphate ions. A higher than normal potassium concentration is associated with kidney failure.

Magnesium is found mainly in the bones, but is also important in the mechanism of muscle contraction, nerve function, several enzymes connected with energy production and as a laxative. It is present in many foods and deficiencies do not occur. Green leafed vegetables contain magnesium as chlorophyll which is metabolized by the body to serve the above purposes. Whole grain contains an appreciable amount of magnesium which is removed in the process of refinement. In general there seem to be few or no advantages associated with the refinement of foodstuffs.

The biological roles of iron are discussed in Chapter 5.

The third group of essential elements consists of chromium, molybdenum, manganese, cobalt, copper, zinc, silicon, selenium, fluorine, and iodine, all of which are present in the body in trace amounts of no more than 0.001% of body weight. Reference to the biological importance of chromium, copper and zinc is made in Chapter 5. Some of the trace elements have only recently been identified as being essential for life because of their minute concentrations. It is difficult to distinguish between an essential element and an impurity element if the concentration is almost that which is just detectable. The elements boron, nickel and arsenic are possible additions to the list of essential trace elements, research on them not being fully conclusive at the present time.

Molybdenum and manganese are essential components of some enzyme systems, manganese being implicated in the development of bones and connective tissue, the metabolism of fatty acids, cholesterol and carbohydrates, and in the maintenance of normal reproductive functions. No deficiencies have been reported, there being sufficient quantities of the two elements in a normal diet.

Cobalt is a constituent of vitamin B_{12} (cobalamin) which is discussed below. Deficiency of cobalt in the form of cobalamin causes the condition known as pernicious anemia.

Selenium is required by the immune system, might assist in the prevention of some cancers and acts as an anti-oxidant to prevent tissue damage. **The protection against some cancers afforded by selenium is not sufficiently proven and should not result in attempts to supplement the normal diet**. This is because there is a fairly narrow margin between the beneficial and toxic levels of the element.

Silicon stimulates growth, hardens teeth and bones, and promotes the formation of connective tissue.

Fluorine, in the form of the aqueous fluoride ion, $F^-(aq)$, is essential for the development and maintenance of teeth, the prevention of dental disease and is thought to prevent the progression of osteoporosis, a disease which causes the embrittlement of bones. It is normally present in drinking water and is currently added to the extent of one part per million of such liquid. Such additions have resulted in a 50-60% reduction of dental caries in children. Some local authorities have been forced, by the democratic action of their electors, to abandon their policy of adding fluoride (as sodium fluoride) to the drinking water in their areas. This has had the effect of restoring the incidence of children's dental disease to its previous poor level. The main function of iodine in the body is to contribute to the production of the thyroid hormone which is vital for metabolism.

9.3 THE MOLECULAR COMPOSITION OF THE HUMAN BODY

The composition of the human body in terms of the types of chemical substances it contains is given in Table 9.9.

Table 9.9 - The composition of the human body by type of molecule

substance	percentage
water	65
protein	18
fat	10
carbohydrate	5
small organic	1
inorganic	1

Proteins are macromolecules which are constructed from amino acid units. The basic structure of an amino acid is shown in Fig. 9.1.

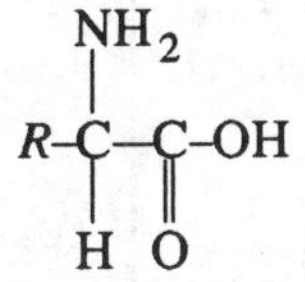

$$\begin{array}{c} NH_2 \\ | \\ R\text{–}C\text{—}C\text{–}OH \\ | \quad \| \\ H \quad O \end{array}$$

Fig. 9.1 The structure of amino acids

There are four groups (*R*, H, NH_2 and COOH) attached to a carbon atom in a tetrahedral manner, the nature of the group *R* characterizing, i.e. being different for, any particular amino acid. There are twenty amino acids which are found in proteins, ten of which the body cannot manufacture and have to be supplied in food. The ten which have to be obtained from food are known as essential amino acids. Their names and abbreviations, which are used later, are given in Table 9.10.

Table 9.10 - The ten essential amino acids

amino acid	abbreviation
arginine	Arg
histidine	His
isoleucine	Ile
leucine	Leu
lysine	Lys
methionine	Met
phenylalanine	Phe
threonine	Thr
tryptophan	Trp
valine	Val

One of the ten essential amino acids, arginine, is essential only for children. Adults synthesize sufficient quantities for themselves. The names and abbreviations of the ten non-essential amino acids are given in Table 9.11.

Table 9.11 - The ten non-essential amino acids

amino acid	abbreviation
alanine	Ala
asparagine	Asn
aspartic acid	Asp
cysteine	Cys
glutamic acid	Glu
glutamine	Gln
glycine	Gly
proline	Pro
serine	Ser
tyrosine	Tyr

Proteins which contain all ten essential amino acids are known as adequate proteins; animal protein, soya bean protein and milk are good examples. Proteins which are deficient in one or more of the essential amino acids are called inadequate proteins. The protein in corn is low in lysine and tryptophan, that in rice is low in lysine and threonine and that in wheat is low in lysine. Vegetarians have to make sure that their food is a suitable mixture of materials which allow a source of all ten essential amino acids, preferably at each meal. Inadequate meals, taken regularly, would lead to protein deficiencies which could threaten health. Protein deficiencies are associated with reduced resistance to disease and sub-normal growth in children. Protein malnutrition is an important and usually fatal disease experienced by third world infants. It accompanies famine and war.

Proteins are constructed from amino acid units by peptide linkages as is shown in Fig. 9.2. The amino group of one amino acid loses one hydrogen atom, the OH group of the COOH group of a second amino acid combining with it to give water, allowing a carbon-nitrogen bond to be made.

The resulting compound is called a dipeptide. The formation of peptide linkages allows the production of many varied chains of polypeptides or proteins. Sometimes the term protein is reserved for polypeptides with more than 20 amino acid units, but there is no other distinction in meaning of the two terms. The cysteine molecule has an *R* group which is CH_2SH and which, if present in two polypeptide chains, allows cross-linking of chains to occur by the elimination of the hydrogen atoms attached to the sulfur atoms:

$$\text{Chain(1)-}CH_2SH + HSCH_2\text{-Chain(2)} \longrightarrow \text{Chain(1)-}CH_2S\text{-}SCH_2\text{-Chain(2)}$$

Cross-linking of this type occurs in the insulin molecule and is described in Chapter 10.

```
     NH2          H R
      |           | |
   R–C––C–O + H–N–C–C–OH
      |  ‖ |      | ‖
      H  O H      H O

           ↓

     NH2  H R
      |   | |
   R–C––C–N–C–C–OH  +  H2O
      |  ‖   | ‖
      H  O   H O
```

Fig. 9.2 The formation of a peptide linkage; the *R* groups may be identical or different

Reactions of polypeptide chains with carbohydrates, lipids (fats), nucleic acid, metal ions and other molecules, produce a great range of proteins which are essential for the structure of the body and its function. Some proteins have chains which coil and fold to give globular molecules which are soluble in water and perform active functions. Others have chains which coil and contribute to fibrous materials that are used for structural and protective purposes. Enzymes and hemoglobin are examples of globular proteins. Enzymes are the molecules which act as catalysts for many reactions occurring in the body. The action of hemoglobin is described in Chapter 5. Fibrous proteins form the basis of hair, nails, skin, bones, teeth, connective tissue, tendons and cartilage.

Carbohydrates form 75% of plants, mainly as cellulose which supports the structure of a plant, and starch which is the energy storage substance. Both are polysaccharides in that they are polymers of sugar molecules. The monomer units from which cellulose and starch are composed are glucose molecules. Glucose, $C_6H_{12}O_6$, is an example of a hexose (because of the six carbon atoms) sugar molecule. There are several isomers of hexoses but only two are of importance for this discussion. The structures of the two isomers of the glucose molecule are shown in Fig. 2.7. They differ only in the positions of the hydrogen atoms and the hydroxyl groups on the first carbon atom. α-glucose is the molecule sometimes known as D-glucose or dextrose. It is a very sweet tasting sugar and is vital to the body's metabolism of foods. It is the molecule which passes into the blood stream to be used for energy production wherever it is needed. The other form, β-glucose is not metabolized by the human body. The commonly used natural sweetening agent, sucrose, is a disaccharide, i.e. it is constituted from two monosaccharide units. The structure of sucrose is shown in Fig. 9.3.

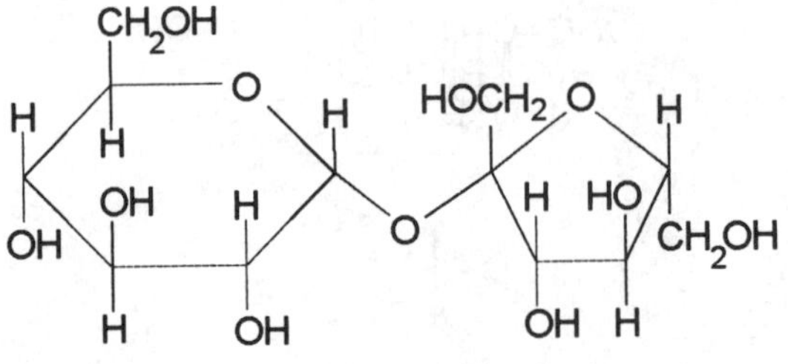

Fig. 9.3 The structure of the sucrose molecule; the element symbols of the carbon atoms in the six and five membered rings are omitted by convention

As is usual with organic molecular structures, most of the carbon atoms are indicated by the junction of four single bonds. When sucrose is hydrolyzed it splits up into the two monosaccharides, α-glucose and fructose, both of which are further metabolized in the body. Starch is a polymer of glucose units which breaks down in stages to give α-glucose when metabolized. Cellulose is a straight chain polymer composed of around 10,000 β-glucose units and since there are no enzymes present in the human body which can catalyse its breakdown it is indigestible. Any cellulose in food ingested by the body passes through the digestive tract as roughage which assists in the elimination of waste material. To that extent it is beneficial to health and should be taken in the diet as bran and whole wheat bread. Starch consists of two different polymeric forms of α-glucose units. One form is that known as amylose which is a straight chain polymer of between 1000-4000 glucose units. The other form is amylopectin which is a branched chain polymer consisting of between 600-6000 units such that there are around fifty units between the branching points. The branched chains are much better for thickening sauces and are present to a greater extent in corn flour than in ordinary flours.

Both forms of starch and any sucrose are hydrolyzed in the body to give the α-glucose monomer molecules. The monomers are converted by the liver and muscles to a polymer called glycogen which has a similar structure to that of amylopectin but which is more highly branched, with about twelve glucose units between branching points. Glycogen represents an energy store in that it is easily hydrolyzed to give glucose when that molecule is required.

Other carbohydrate molecules are incorporated with fat and protein molecules for special purposes in the body, particularly in the composition of mucous and lubricating fluids, cartilage, bone and in the membranes of red blood cells.

Fats are compounds of glycerol with acids which have long hydrocarbon chains. The glycerol molecule has the formula $CH_2OH.CHOH.CH_2OH$ (see Fig. 9.4) and possesses three hydroxyl groups which can all interact with the COOH groups of acid molecules to give compounds known as triglycerides, such molecules representing the bulk of the fats in living beings. Monoglycerides and diglycerides, produced by the reactions of glycerol with either one or two fatty acids, are contributors to the body's fat in smaller fractions than are triglycerides. The main long-chain acids which contribute to human fat composition are:

myristic acid; $CH_3(CH_2)_{12}COOH$
palmitic acid; $CH_3(CH_2)_{14}COOH$
stearic acid; $CH_3(CH_2)_{16}COOH$
oleic acid; $CH_3(CH_2)_7CH{=}CH(CH_2)_7COOH$
linoleic acid; $CH_3(CH_2)_4CH{=}CHCH_2CH{=}CH(CH_2)_7COOH$

The first three acids are saturated, i.e. they possess no double C=C bonds, oleic acid possessing one C=C double bond and linoleic acid having two. Oleic and linoleic acids are unsaturated. They have the capacity to react with dihydrogen so that every carbon atom participates in four single covalent bonds. The manner in which three of these acids combine with glycerol to form a fat is demonstrated in Fig. 9.4.

$$
\begin{array}{lcl}
R.\text{CO.OH} + \text{HO–}CH_2 & & R.\text{CO.O–}CH_2 \\
R.\text{CO.OH} + \text{HO–}CH & \longrightarrow & R.\text{CO.O–}CH \quad + \; 3H_2O \\
R.\text{CO.OH} + \text{HO–}CH_2 & & R.\text{CO.O–}CH_2
\end{array}
$$

3 acid molecules + glycerol ⟶ fat + 3 water molecules

Fig. 9.4 The formation of a triglyceride fat; the R groups may be identical or different

Another possible combination is that between glycerol, two of the acids and one phosphate group to give a phospholipid. Many kinds of phospholipids are found in the body. All the fats represent an energy store and also contribute to the protection of the structure of the body against accidental damage, e.g. in physical shock absorption, and to maintain cell membranes, skin and other tissues.

Some of the small molecules found in the human body are the well known compounds called vitamins and hormones. Vitamins are essential for the maintenance of health and either cannot be synthesized by the body or are not synthesized in sufficient amounts. They must be present in a well balanced diet or taken in tablet form. Chemically, they are described as being co-enzymes which are compounds which assist enzymes in their catalytic actions. They are required in small quantities and are classified in terms of their solubility or otherwise in water.

The water soluble vitamins occur in meats, liver, kidney, whole grains, leafy vegetables, milk, eggs, tomatoes, potatoes, legumes and citrus fruits.

Vitamin C, or ascorbic acid, is an agent in the synthesis of collagen and in amino acid metabolism. Linus Pauling has recommended 'megadoses' of vitamin C to ward off common colds and other diseases. There is some evidence that doses of 500 milligrams per day prevent the occurrence of the common cold. In the author's self-experimentation over the last ten years such dosage does have a positive effect. What the state of the author's health would have been over the last ten years had he not taken daily doses of vitamin C will never be known. Vitamin B_1, or thiamine, contributes to the metabolism of carbohydrate as a co-enzyme. Co-enzymes are compounds which assist enzymes in their catalytic activities. Vitamin B_2 (chemical name: riboflavin) and niacin (nicotinic acid) are co-enzymes in oxidation reactions. Vitamin B_6, (pyridoxine) is a co-enzyme for the metabolism of amino acids and fatty acids. Vitamin B_{12}, or cobalamin, is a co-enzyme in the metabolism of nucleic, amino and fatty acids. Its molecule contains a central cobalt atom. It is essential for the maturation of red cells in the bone marrow, deficiency causing pernicious anemia. It functions in close co-operation with folic acid, a deficit of which also causes an identical anemia. Pantothenic acid is part of a centrally important co-enzyme called Co-A which participates in major metabolic processes of the digestion of proteins, carbohydrates and fats and in the synthesis of those classes of compounds for use in the body. Folacin, or folic acid, is a co-enzyme in nucleic acid and amino acid metabolism. Biotin forms part of the enzymes which are used in the metabolism of carbohydrates and fats.

The water insoluble vitamins are, because of their non-polar nature, soluble in fat and can exist in the lipid sections of membranes. They are found in green and yellow vegetables, fish and vegetable oils, eggs and dairy products.

Vitamin A is important in the formation of eye pigments, in the maintenance of mucous membranes and in the transport of nutrients through cell membranes.

Vitamin D is important in the regulation of calcium and phosphate metabolism.

Vitamin E, or tocopherol, is concerned with the maintenance of cell membranes.

Vitamin K is important in the synthesis of blood clotting agents in the liver.

Any person enjoying a balanced diet of normal food is assured of a sufficient supply of vitamins. There are many formulations of mixtures of vitamins available at the pharmacy for those people who do not eat well or think that their diet is deficient in some way.

Hormones are a group of chemical messengers which control the activities of cells, tissues and organs of the body. They are produced in various glands and are transported around the body in the blood stream and other body fluids. Some hormones are steroid molecules which are non-polar and can pass through cell membranes. Others are peptide molecules which are polar and cannot pass from the blood into the interior of cells. The hormones have specific functions as the following examples demonstrate.

Thyroxine, whose structure is shown in Fig. 9.5, is an amino acid produced in the thyroid gland and which controls metabolic rate and regulates growth. The formulation of the hexagonal C_6 benzene rings is dealt with in detail in Chapter 11. The main point of the diagram is to indicate the presence of the four atoms of iodine in the thyroxine molecule. To maintain a sufficient concentration of thyroxine in the body requires a sufficiency of iodine in the diet and emphasizes the essentiality of the element.

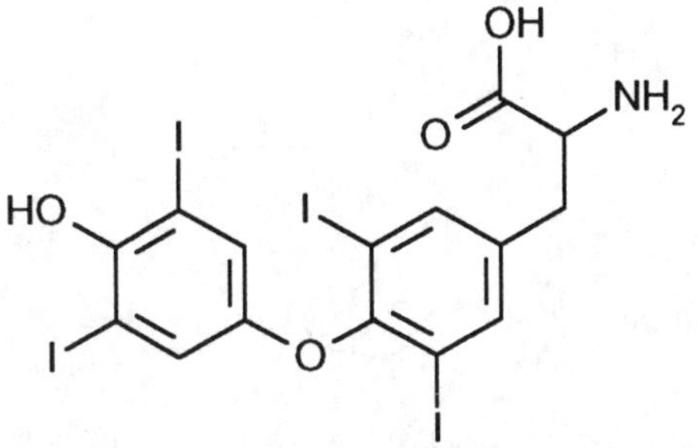

Fig. 9.5 The structure of the thyroxine molecule; two hydrogen atoms are omitted from the two C_6 planar hexagonal benzene rings

A pair of peptide protein hormones, insulin and glucagon, which are produced in the pancreas control the level of glucose in the blood. If the glucose level increases beyond a certain level the intervention of insulin causes the liver and muscles to manufacture glycogen and so reduce the blood sugar level. There is evidence for the essential participation of chromium as a co-factor or potentiator of insulin known as the glucose tolerance factor; this is described in more detail in Chapter 5. A deficiency of chromium can cause the late-onset form of diabetes which afflicts some people as they reach an advanced age. Although most of the roughage foodstuffs contain acceptable amounts of chromium the best, and most palatable, source is wine. It would seem good preventive medicine to have a daily intake of such liquids. If the glucose level becomes too low, the pancreas releases glucagon which stimulates the liver to convert glycogen into glucose and so increases the blood sugar level. These interactions are shown in the diagram of Fig. 9.6.

$$\text{glucose} \underset{\text{glucagon}}{\overset{\text{insulin}}{\rightleftharpoons}} \text{glycogen} \rightleftharpoons \text{fat}$$

Fig. 9.6 Interactions between glucose, glycogen and fat

If, as the result of overeating, the glycogen stores become full glycogen is converted into the long term energy storage material, fat, as is also shown in Fig. 9.6. Slimmers may have noticed that they very quickly put on 1-2 lb (0.5-1 kg) of weight if they relax their diet. This is because their slimming routine depletes their glycogen store and even one 'full' meal is sufficient to replenish it. If, as a result of overeating, slimmers put on more weight than 2lb (1 kg) that weight is likely to be in the form of the more permanent storage of energy, i.e. fat.

The steroid hormones are fat-soluble molecules with structures based upon that of the alcohol known as cholesterol whose structure is shown in Fig. 9.7, together with those of the related molecules of cortisone, testosterone, estrone and progesterone.

cholesterol
HO
cortisone
OH
O
O
OH
O
testosterone
OH
O
estrone
O
HO
progesterone
O
O

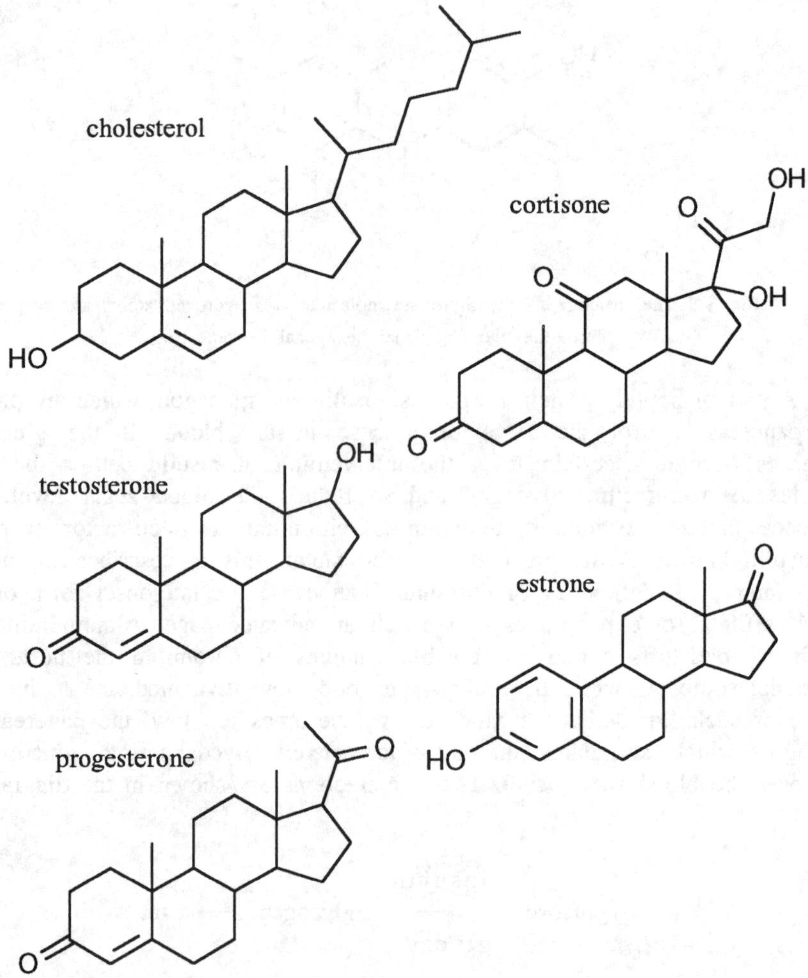

Fig. 9.7 The structures (in descending order) of cholesterol, cortisone, testosterone, estrone and progesterone; they are shown as carbon skeletons with omission of all the hydrogen atoms connected to atoms of carbon; they share the same basic skeleton, the relatively minor structural differences being responsible for their different functions in the body

The -ol ending of cholesterol indicates that the molecule contains an -OH group, the -one endings of the other molecules in Fig. 9.7 arises from the C=O groups which they contain. The quite similar structures are shown as carbon skeletons as is usual for large organic molecules. Also omitted from the diagrams are the hydrogen atoms which are attached to the carbon atoms, in each case there being a sufficient number to satisfy the tetravalency of the carbon atoms.

Cortisone is essential for carbohydrate metabolism and is also used as a drug in the treatment of rheumatoid arthritis.

Testosterone is the male sex hormone and regulates the development of the male reproductive organs. Estrone is the female sex hormone.

Progesterone is a sex hormone produced in pregnancy to prepare the uterine lining for the embryo.

Cholesterol is synthesized in the liver and is converted to the steroid hormones when necessary. It also forms part of cell membranes, the synthetic substance and any that is present in the diet being transported to required sites by the blood. The molecule circulates in association with various forms of what are generally known as low density lipoproteins (LDL), which are combinations of protein and fat molecules. Excesses of cholesterol are removed from cell surfaces by high density lipoproteins (HDL) and returned to the liver. High levels of LDLs and low levels of HDLs in the blood correlate with a higher incidence of heart disease, strokes and atherosclerotic vascular disease. The average cholesterol concentration in the blood is 225 milligrams per litre, with values of 200 and greater than 300 being associated with moderate and high cardiac risks respectively. There are risks associated with cholesterol reducing drugs, the better treatment for people with higher than normal values being diet adjustment. Diet alone may not produce the necessary reduction in cholesterol level, in such cases tablet therapy becoming mandatory.

The essential trace elements are important constituents of some enzymes which catalyse the many processes occurring in the body. These processes collectively represent the metabolism of the body. Metabolism comprises the changes occurring to the food intake and the processes of breaking down of large molecules, which is known as catabolism, and of building large molecules, which is called anabolism. The cooking of food before consumption causes some breakdown of the macromolecules which makes the body's task easier and the food to be tastier. Carbohydrate metabolism starts in the mouth with the processes of chewing and partial breakdown of the long starch molecules into shorter chains called dextrins and into the disaccharide maltose. The process is catalysed by enzymes in the saliva. The partial breakdown products pass through the stomach without much change and are finally broken down into glucose in the small intestine. The glucose passes through the membrane of the intestine and joins the blood stream where its level is controlled as described above. The glucose which is not immediately used by the muscles to provide energy is stored initially as glycogen and ultimately as fat if the glycogen storage is full.

Protein metabolism occurs mainly in the highly acidic conditions of the stomach where the macromolecules are broken down to give short chain peptides and individual amino acids. The process is completed in the small intestine with some of the small peptides passing into the blood, the remainder being completely broken down to their constituent amino acids. The amino acids are then transported to the muscles where they are used to build and maintain the constituent proteins of the muscles. The building of muscle tissue is aided by anabolic steroids which are chemically similar to testosterone but which have been designed to be taken orally. They have unpleasant and dangerous side-effects, but are taken by some athletes to cheat their opposition. Any excess of protein in the diet may be used for energy production, the nitrogen content being converted to ammonia and then to urea which is excreted in the urine. Starving people who are using their body protein for energy often excrete ammonia as the ammonium ion because the process of conversion of ammonia to urea is too slow for the amount of protein breakdown.

Fat metabolism starts in the stomach and is completed in the small intestine, the macromolecules breaking down into fatty acids, glycerol and monoacylglycerols. These substances are then used to construct molecules for use in membranes and tissues, any excesses being converted into fat. The energy stored in the form of fat can be released when necessary by the fat molecules being degraded and converted into glucose.

In summary, the possible inter-conversions of the substances in foods to those in the body are given in Table 9.12.

Table 9.12 - Metabolic conversions of constituents of food to constituents of the body

food constituent	body constituents
carbohydrates	carbohydrates and fat
fats	carbohydrates and fat
proteins	carbohydrates, proteins and fat

To avoid having an excessive fat content, a person should have a diet which has sufficient carbohydrate to provide the energy requirements of the body and optimum amounts of fat and protein for the building and maintenance of the body's organs, membranes, fluids and structures. Overweight people should eat a diet which consists of less than their daily needs of carbohydrate and fat but which has sufficient protein to ensure that none of their body protein is used up. To avoid the loss of protein in such circumstances it is essential to participate in exercise to maintain muscles and to ensure that the extra energy requirements are met by the use of body fat.

9.4 THE CHEMISTRY OF LIFE

The chemistry of human reproduction and genetics is very complicated but, if the details of the many enzymes, which are protein molecules which act as catalysts for specific reactions, are omitted, there is a great simplicity in the subject. The human genome is the famous double helix molecule of deoxyribonucleic acid (DNA) with a relative molar mass of around three and a half billion (3,500,000,000).

A single strand of DNA consists of many molecules of a derivative of ribose (a 5-carbon sugar) known as deoxyribose which are linked together with phosphate (PO_4) groups. Attached to each of the deoxyribose molecules is a molecule of one of four base substances. The structures of ribose and its derivative, deoxyribose, are shown in Fig. 9.8 together with the conventional numbering of the carbon atoms.

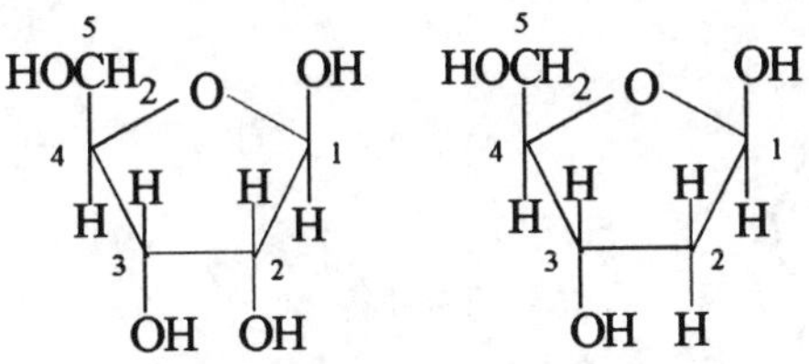

5 $HOCH_2$ O OH — 4, 1 — H H H H — 3, 2 — OH OH

ribose

5 $HOCH_2$ O OH — 4, 1 — H H H H — 3, 2 — OH H

deoxyribose

Fig. 9.8 The structures of ribose and deoxyribose; the conventional numbering of the carbon atoms is shown, hydrogen atoms of the OH groups of atoms 3 and 5 being displaced in reactions with phosphate ions in RNA and DNA formation

The molecules of deoxyribose in DNA are connected by what are known as phosphate ester linkages in the 3 and 5 positions. The formation of the ester linkages may be formulated in terms of the reaction of phosphoric acid, H_3PO_4, which may be written as $O{=}P(OH)_3$ (because it contains three P-OH linkages), with the OH groups of two deoxyribose molecules, which may be represented by the formula HO-(D)-OH, the particular OH groups being those on carbon atoms 3 and 5.

$$\text{HO-(D)-OH} + \text{HO-}\overset{\overset{\displaystyle O}{\|}}{\underset{\underset{\displaystyle OH}{|}}{P}}\text{-OH} + \text{HO-(D)-OH} \longrightarrow \text{HO-(D)-O-}\overset{\overset{\displaystyle O}{\|}}{\underset{\underset{\displaystyle OH}{|}}{P}}\text{-O-(D)-OH} + 2H_2O$$

In DNA, the hydrogen atom of the hydroxyl group on carbon number 1 of each deoxyribose molecule is replaced with one of the four bases; adenine (A), thymine (T), cytosine (C) or guanine (G), their structures being shown in Fig. 9.9, together with that of uracil (U) which is used instead of thymine in the formation of RNA, which is discussed below.

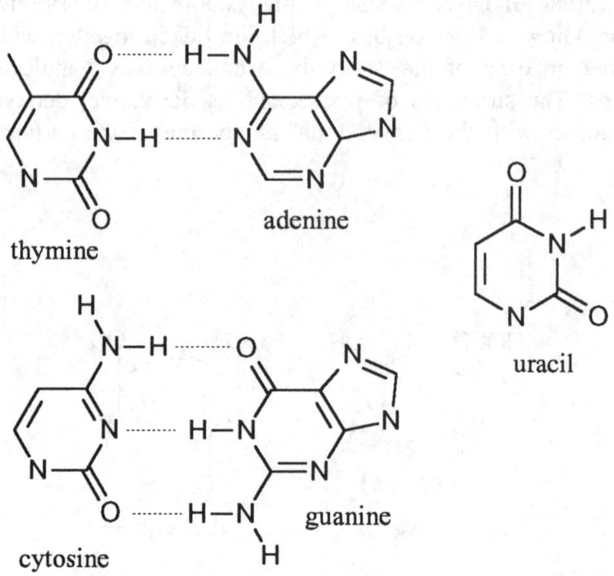

Fig. 9.9 The structures of the bases adenine (A), thymine (T), cytosine (C), guanine (G) and uracil (U); the hydrogen bonds linking (T) with (A) and (C) with (G) in DNA are indicated by the dotted lines

The arrangement of deoxyribose (D), phosphate (P) and base units (A, T, C or G) is shown in Fig.9.10 as a straight chain.

```
      A   T   C   G
      |   |   |   |
....-D-P-D-P-D-P-D-P-....
```

Fig. 9.10 An arrangement of units in a single DNA chain

The arrangement of two such chains is shown in Fig. 9.11, the method of bonding being the formation of hydrogen bonds between one or two hydrogen atoms on one of the base pairs and electronegative nitrogen or oxygen atoms on the other base, shown by the dotted lines in the diagram.

```
....-D-P-D-P-D-P-D-P-D-P-....
     |   |   |   |   |
     A   C   T   G   C
     :   :   :   :   :
     T   G   A   C   G
     |   |   |   |   |
....-D-P-D-P-D-P-D-P-D-P-....
```

Fig. 9.11 Two cross-linked chains in the DNA structure

In order to maximize the hydrogen bonding interactions between the two strands and to prevent the adjacent bases from impeding each other it is necessary for them to adopt the double helix structure as shown in the diagram of Fig. 9.12.

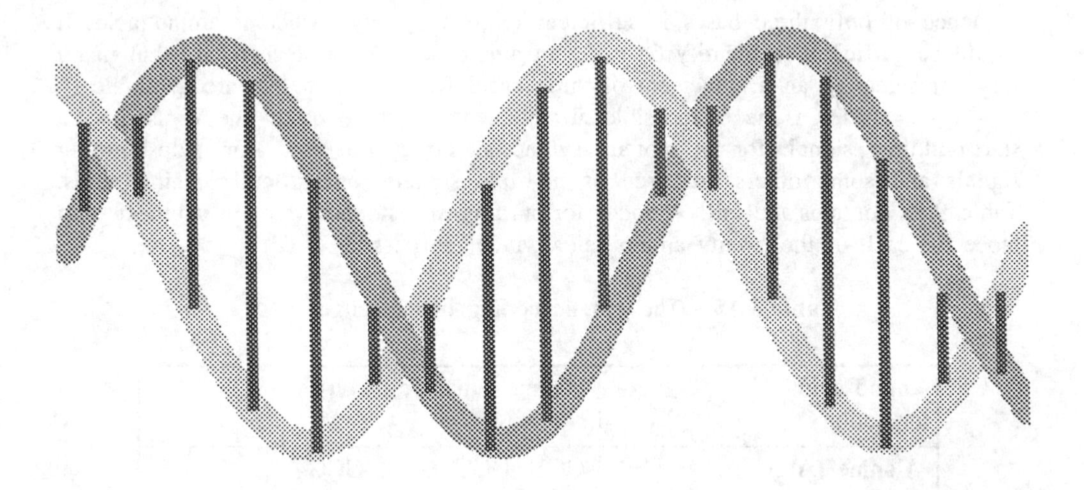

Fig. 9.12 A diagram of a portion of a DNA chain showing the double helix structure. The cross-linkages between the two ....sugar-phosphate-sugar.... chains consist of cytosine-thymine (C-T in Fig. 9.9) or adenine-guanine (A-G in Fig. 9.9) base pairs hydrogen bonded together

The constraint placed upon the DNA structure by the four bases and the need to maximize the hydrogen bonding between them is to restrict the base pairs (those that face each other across the two strands) to adenine-thymine (A-T) and cytosine-guanine (C-G). These two arrangements are indicated in Fig. 9.11 with the hydrogen bonds represented by dotted lines. The details of the hydrogen bonding between adenine and thymine and between cytosine and guanine are shown by the dotted lines in Fig. 9.9. Other pairings are not as satisfactory and this restriction introduces a simplification of this feature of the DNA structure. The particular sequence of bases on the DNA strands contains all the necessary information for the self-replication of the DNA double helix and for the construction of every protein molecule needed by the body in its lifetime. As the parent DNA double helix unwinds at one end the two strands acquire their new complementary partners gradually as the sugar phosphate and base ingredients are used. The two daughter double helixes are exact copies of their parent helix. The genetic code which is used to convey protein synthesis information to the ribosomes in a cell where the proteins are constructed is remarkably simple. A portion of the DNA helix unwinds and allows a molecule called messenger RNA (mRNA) to be formed by matching its bases with their complements on the DNA chain. Ribonucleic acids (RNAs) are constructed in a similar manner to that of DNA except that they usually are single strands, they use ribose (the structure of which is shown in Fig. 9.8) instead of deoxyribose and the four bases are cytosine (C), guanine (G), adenine (A) and uracil (U), the last being used in place of thymine (T) and substituting for it on every occasion. The structure of uracil is shown in Fig. 9.9.

There are twenty amino acids which may be used as protein building blocks and four bases which may be used to signify the identity of any one amino acid. A sequence of only three bases is sufficient to identify any particular amino acid. It would be sufficient to identify 64 different amino acids because there are that many ways of selecting an arrangement of three bases from the four that are available (4 x 4 x 4 = 64). Of the 64 possible arrangements of three bases one is used as a start-building signal for a protein synthesis, three are used for stop-building signals and some others are used as multiple signals for particular amino acids. Table 9.13 contains the mRNA codes for starting and stopping protein synthesis and those for each of the twenty amino acids which are possibly used.

Table 9.13 - The genetic coding in messenger RNA

amino acid or action	coding sequence of bases
Alanine (Ala)	GCU, GCC, GCA, GCG
Arginine (Arg)	CGU, CGC, CGA, CGG, AGA, AGG
Asparagine (Asn)	AAU, AAC
Aspartic acid (Asp)	GAU, GAC
Cysteine (Cys)	UGU, UGC
Glutamine (Gln)	CAA, CAG
Glutamic acid (Glu)	GAA, GAG
Glycine (Gly)	GGU, GGC, GGA, GGG
Histidine (His)	CAU, CAC
Isoleucine (Ile)	AUU, AUC, AUA
Leucine (Leu)	UUA, UUG, CUU, CUC, CUA, CUG
Lysine (Lys)	AAA, AAG
Methionine, (Met), START	AUG
Phenylalanine (Phe)	UUU, UUC
Proline (Pro)	CCU, CCC, CCA, CCG
Serine (Ser)	UCU, UCC, UCA, UCG, AGU, AGC
STOP	UAA, UAG, UGA
Threonine (Thr)	ACU, ACC, ACA, ACG
Tryptophan (Trp)	UGG
Tyrosine (Tyr)	UAU, UAC
Valine (Val)	GUU, GUC, GUA, GUG

As a trivial example of the synthesis of a protein with seven amino acid units the information on the DNA strand, grouped in threes for convenience, might be:

TAC TTT CGA GTA CCA CAA AAT ATC

which would transcribe onto the messenger RNA in the order shown in Table 9.14:

Table 9.14 - The transcription of information from DNA to mRNA and the production of protein

DNA code	mRNA code	amino acid unit
TAC	AUG	Met
TTT	AAA	Lys
CGA	GCU	Ala
GTA	CAU	His
CCA	GGU	Gly
CAA	GUU	Val
AAT	UUA	Leu
ATC	UAG	-

The AUG code is the signal for the start of the protein to be made and its initial amino acid, methionine. The protein which is then constructed has the sequence:

Met-Lys-Ala-His-Gly-Val-Leu

since the code UAG indicates the end of building. The actual proteins produced in the body vary greatly in size and can have as many as several thousand amino acid units. The hormones which control the glucose levels in the blood, insulin and glucagon, are relatively small proteins which contain 51 and 21 amino acid units respectively. There are 574 amino acid units in the hemoglobin protein.

So far, only the primary structure has been described; that of the amino acid sequence. The secondary structure is that of the primary chain as it forms a helix as the result of hydrogen bonding interactions. The tertiary structure arises as the result of the coiling of the helix onto itself to produce an apparently disordered arrangement. There is order in such tertiary structures in that they are the same for a given protein, and the various cavities offer receptor sites for molecules whose reactions would otherwise be many factors of ten slower. Some proteins form a quaternary structure, hemoglobin being a good example. In hemoglobin four of the coiled tertiary globin proteins, each with its heme molecule containing an iron atom, agglomerate to produce an oxygen carrier with a capacity of four dioxygen molecules per hemoglobin protein. As is described in detail in Chapter 5, the dioxygen is carried through the blood to muscular sites and transferred to the static myoglobin proteins which store the dioxygen until needed by the muscles to provide energy by glucose oxidation.

The DNA molecule is the source of the information which is required to synthesize a protein required by the body. The DNA is used to make a template molecule called messenger RNA (mRNA) which is used in protein construction. The proteins of the body carry out their built-in duties of controlling growth, maintaining cells, catalysing metabolism and storing energy. Retro-viruses, e.g. HIV which causes the disease known as acquired immune deficiency syndrome, abbreviated as AIDS, exist which have the capacity, using their own RNA, to alter the parent cell's DNA by a process of reverse transcription. The altered DNA is then replicated by the normal process to the eventual detriment of the host's health.

The function of DNA is summarized by the diagram shown in Fig. 9.13.

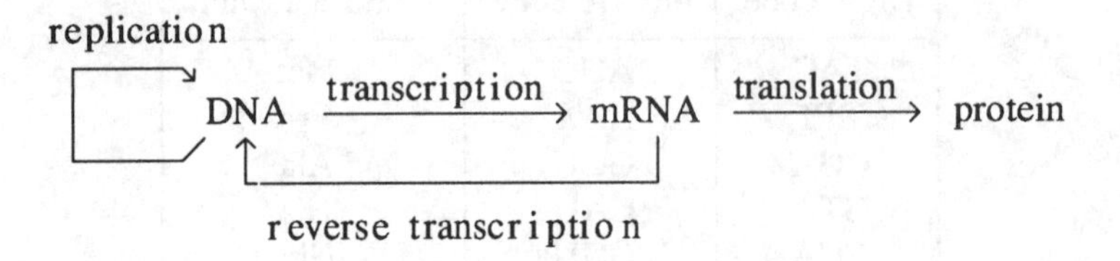

Fig. 9.13 The basis of DNA function

Human DNA consists of six thousand million base pairs which indicates that it would have a relative molar mass of about 3.5 million million. It is not a single molecule however, being split up between forty six chromosomes in the nucleus of each human cell in the form of twenty three pairs. A male cell contains twenty two pairs of chromosomes in which each of the partners has a similar appearance. The other pair are dissimilar in shape, the larger one being called the X chromosome, because it is roughly X-shaped, as are the remaining 44, the small one being termed the Y chromosome; its shape is vaguely that of the letter Y. A female cell contains two X chromosomes in addition to the 44 which are similar to those in the male cell. The X and Y chromosomes determine the gender of any offspring resulting from the merging of a male cell with a female cell. When a cell divides in the course of normal growth within an individual the nucleus of the cell divides so that two nuclei are produced, both with a full complement of 46 chromosomes. In the process of division, sections of the DNA are interchanged between the paired chromosomes so that inherited characteristics of both parents become intermingled. Sperm and egg cells which are used in sexual reproduction differ from the normal body cells in that they each possess only one chromosome from each of the pairs in the body cells. There are 2^{23} (twenty three 2s multiplied together) = 8,388,608 ways of making a selection of two chromosomes twenty three times. The process of selection is done twice, once in the sperm cell and once in the egg cell, the combination of the two selections producing the 46 chromosomes of the new human being. The number of ways of producing the new human cell is 8,388,608 x 2 = 16,777,216, so it is not surprising that each human being is unique. To pay someone the compliment of being '*one in a million*' does not sound so good any more!

The sperm cell contains the selecting power for the gender of the new human in that it may contribute either the X chromosome or the Y chromosome, the egg cell inevitably contributing an X chromosome to the combined cell from which the new human develops. The cell of the new human produced by the fusion of the sperm and egg cells contains its complement of 46 chromosomes, with the possibilities of its containing XY (male) or XX (female) combinations of the so-called sex chromosomes:

sperm (X) + egg (X) $\longrightarrow$ female (XX)

sperm (Y) + egg (X) $\longrightarrow$ male (XY)

The many other characteristics of a human being are similarly handed down by the same selection process from the parent cells. The genes that govern the particular and detailed development of a human are sections of the DNA chains of the chromosomes, the genetics having been determined in the selection of the twenty three components of the sperm and egg cells from their individual sets of forty six chromosomes. Some characteristics are determined by a single pair of genes, one on each of the paired chromosomes, while others are affected by a number of gene pairs. DNA 'fingerprinting' is now possible. Because every person has a different DNA sequence of bases, it is theoretically, and now practically, possible to identify a sufficiently long base sequence to distinguish one person from another. In addition, it is possible to come to positive conclusions as to whether or not a particular person is the parent of a given child. The father and mother each donate one half of their DNA to their child.

10

Drugs and health

Drugs and poisons. Value of production of pharmaceuticals. Changes in life expectancy. Causes of death. Infectious and non-infectious diseases. Diabetes and insulin manufacture. Treatment of peptic ulcers. Aspirin. Treatment of depression. Improvements in the health of humans.

10.1 DRUGS, LIFE EXPECTANCY AND CAUSES OF DEATH

Drugs are substances which are used in the prevention and treatment of diseases. Such substances are generally beneficial when used in their optimal doses, but which become detrimental, or even poisonous, when used in excessive amounts.

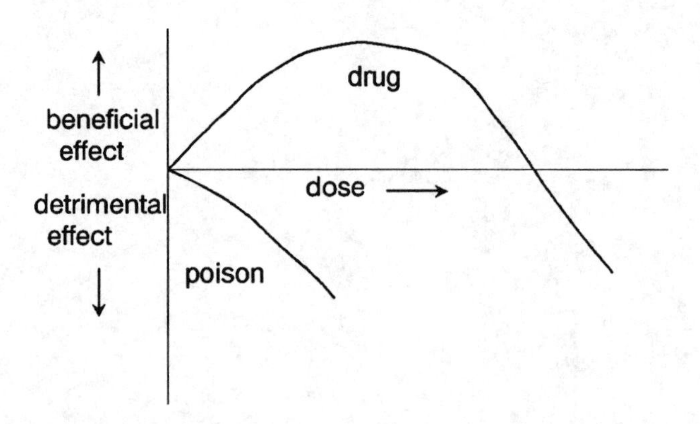

Fig. 10.1 A diagram showing the difference between the effects of drugs and poisons

The diagram in Fig. 10.1 shows the difference between the effects of drugs and poisons on the living organism. Low to medium doses of appropriate drugs have a beneficial effect, but overdoses are usually detrimental. Any dose of a poison is detrimental. The pharmaceutical industry produced drugs worth $140 billion in 1990. The total value of the production of the ten major metals described in Chapter 9 amounted to $116 billion in the same year, placing in perspective the importance to the human race of its desire for good health.

One method of assessing the value of the effectiveness of drugs upon the health of the human race is to compare life expectancies at different periods of history. For example, the percentages of males surviving to the age of sixty who were born at various times since 1690 (in England) are shown in Table 10.1. Such an impressive improvement in life expectancy cannot be ascribed entirely to better medicine.

Table 10.1 - The percentage of males reaching an age of sixty at various periods

period	% surviving to 60 years
1693	19.1
1846	35.6
1895	41.0
1921	58.8
1971	80.2

Other factors which have an input to such figures are better nutrition and general hygiene and the wider use of water purification and sewage systems.

Another indication of the efficacy of drug use is the reduction in annual death rate from infectious diseases which in the mid-nineteenth century was around 11 per thousand and is currently below one per thousand. Deaths from non-infectious diseases have remained fairly constant at between 10-12 per thousand over the same time span. The reductions in the death rates from particular diseases have been brought about by the identification of the causal microorganisms and viruses, the invention and production of antibiotics, antiviral substances and vaccines. During the twentieth century the diseases, tuberculosis, typhus and other fevers, scarlet fever, dysentery, smallpox, diphtheria, whooping cough, poliomyelitis and tetanus, have largely been eliminated by the production of appropriate and effective vaccines. The major causes of death are now those given in Table 10.2.

Ischaemic heart disease is produced by atheroma, an active disease of the artery wall. It is not the simple collection of cholesterol which has diffused through the arterial lining. Atheroma narrows the coronary artery and produces the condition known as angina pectoris. If the lining ulcerates and a clot forms there may be a rapid occlusion of the vessel and a heart attack occurs. Cerebrovascular disease leads to strokes which are cerebrovascular accidents which include thrombosis of a brain artery, embolism (obstruction by a clot from somewhere else in the circulatory system) and cerebral brain hemorrhage, the latter event being less common, but more usually fatal.

Table 10.2 - Major causes of death
England and Wales; all ages, 1980
Total annual death rate = 12 per thousand

cause of death	% of death rate
ischaemic heart disease	26.5
malignant neoplasm (cancer)	22.5
cerebrovascular disease	12.3
other heart conditions	6.8
bronchitis, emphysema, asthma	4.4
accidents, poisoning, violence	3.5
others	24.0

Apart from the prolongation of life, the application of medicine in easing the symptoms of fatal diseases and the treatment of the many non-fatal diseases has led to a significant increase in the quality of life generally. It is very difficult to measure this increase, each disease having its own specific history.

In this chapter, specific discussion of disease is restricted to the examples of diabetes, peptic ulcers and conditions which benefit by treatment with aspirin.

10.2 DIABETES MELLITUS

Diabetes mellitus was a fatal disease before 1920. The discovery that insulin production in the pancreas was essential to the metabolism of glucose was made by Schafer in 1916. The isolation of insulin was made by Banting and Best in 1922, the amino acid sequencing of the molecule being carried out by Sanger in 1955. A lack of insulin causes blood sugar levels which are too high and produces excretion of glucose in the urine. After the discovery that bovine insulin injected into the blood streams of diabetic patients allowed the control of blood glucose levels, the disease became manageable. Although people do still die from the disease, the death rate is fairly low at around 125 per million, death occurring usually when the control of blood sugar levels is lost either by mismanagement, by accident or the loss of control over the many factors that affect the course of the disease. The numerical value of the death rate from diabetes is doubtful because many diabetics die from arterial degeneration and the cause of death is recorded conventionally as disease of the heart, brain or kidney.

The amino acid sequence in human insulin is shown in Fig. 10.2. The A and B chains are joined together by two sulfur-sulfur bonds. Before 1985, all diabetic patients who needed insulin injections used bovine insulin. Human insulin has been produced by four successive processes. The first method was to modify porcine insulin chemically by replacing the terminal alanine unit of the B-chain with threonine.

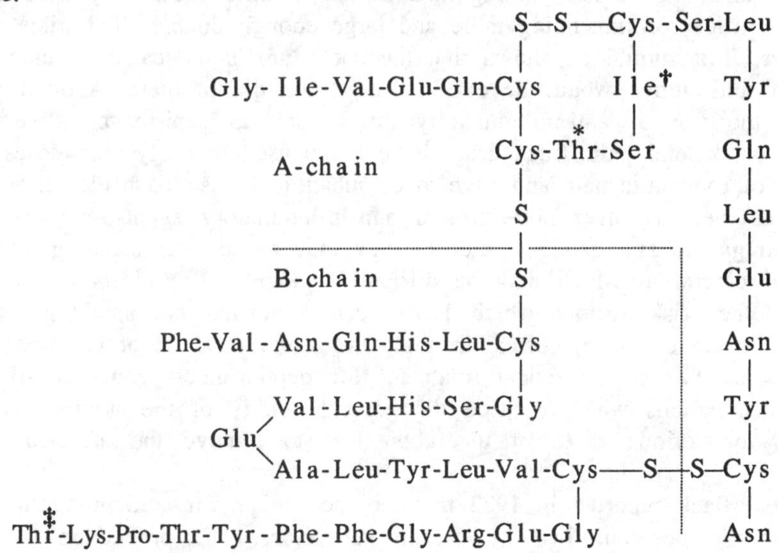

Fig.10.2 The amino acid sequence in human insulin;
the A and B chains are separated by the dotted line
*replaced by Ala in bovine insulin
†replaced by Val in bovine insulin
‡replaced by Ala in bovine and porcine insulin

The other three processes make use of what is known as recombinant DNA technology in which the strand of human DNA containing the information for the construction of a particular protein, e.g. insulin, is spliced into the DNA of a monocellular organism such as yeast or the bacterium *Escherichia coli*, abbreviated to *E. coli*. One method produced the A and B chains in separate processes using altered *E. coli* cells, the two chains being allowed to combine in a subsequent stage. Another used yeast cells to produce proinsulin, a larger protein than insulin, from which insulin was then made. Proinsulin contains a thirty unit chain of amino acids which, via peptide linkages, takes the A-chain (not the A train) - the glycine end - and forms a bridge to the threonine end of the B-chain. The latest process makes use of proinsulin synthesized by altered *E. coli* cells. By 1989, 75% of British diabetic patients were using human insulin to control their disease, the proportion being expected to increase further as more patients try and become accustomed to the natural substance after years of treatment by bovine insulin.

10.3 PEPTIC ULCERS

Another disease whose treatment has changed remarkably in recent years is the peptic ulcer or stomach ulcer. The origins of peptic ulcers are not certain. The organism *Heliobacter (Campylobacter) pylori* infects the duodenal mucosa and could cause ulceration. It is thought to be the definite cause of gastritis and a possible cause of gastric ulcers. It is not yet regarded as the definite cause of duodenal ulcers, some authorities recommending more and large enough double-blind trials to settle the matter. If it should be shown that the bacterium is the cause of ulceration, a course of anti-biotics would be expected to solve the problem. Another possible cause is the use of anti-inflammatory drugs such as aspirin to relive chronic muscular or skeletal pain. Such drugs have to be used in fairly high doses for the treatment of rheumatic pain and have to be substituted if stomach ulceration occurs. Aspirin, as well as other non-steroidal anti-inflammatory agents, may re-activate old healed gastric and duodenal ulcers. They may cause fatal bleeding by causing superficial ulceration which may be difficult to locate. The ulcers rarely become chronic. Other risk factors which have been identified are smoking, excessive alcoholic beverage intake, coffee drinking and acute physical or emotional stress. Before 1980, the only chemical treatment for peptic ulcers consisted of various antacid formulations which reduced the general acidity of the stomach. The sole treatment for serious cases of ulceration was to remove the affected area by surgery.

James Black reported in 1972 that the mechanism for acid production in the stomach could be controlled by certain substances. The production of acid is stimulated by various molecules, including one called histamine whose structure is shown in Fig. 10.3.

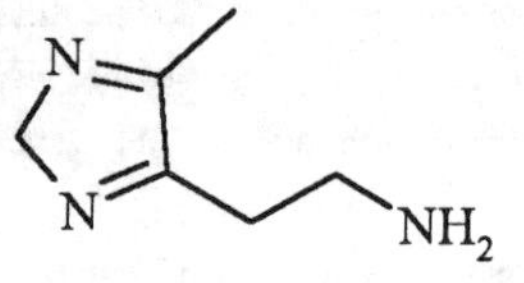

N
N
NH_2

Fig 10.3 The structure of histamine

Such molecules are known as agonists, their action being reduced by antagonists, the drugs which were designed as a result of Black's work (he was awarded the Nobel Prize for Medicine in 1988). The reduction of acid secretion in the stomach by suitably designed molecules transformed the treatment of peptic ulcers. The first successful peptic ulcer drug was cimetidine (Tagamet™) whose structure is shown in Fig. 10.4. It is still the first drug to be tried on patients.

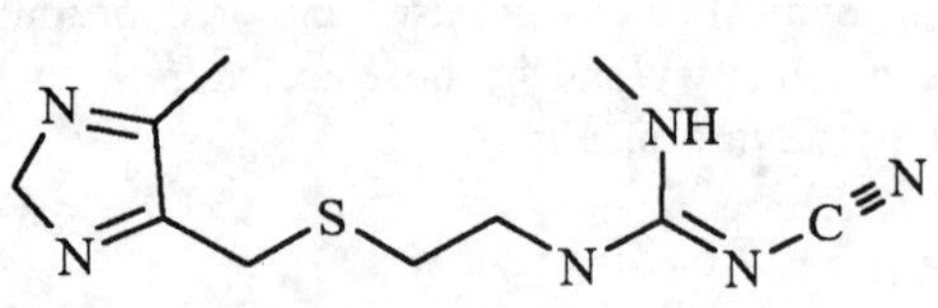

N
NH
N
S
N
N
C≡N

Fig. 10.4 The structure of cimetidine (Tagamet™)

The theory of the drug action is that the five-membered ring of the histamine molecule fits into a receptor site in the cell and stimulates the production of the acid which is necessary for food metabolism. If there is an ulcer present the amount of acid production can be regulated by allowing the cimetidine molecule to take the place of the histamine molecule in a proportion of the acid producing cells. The five-membered ring is accepted by the receptor site but the long side chain somehow prevents the cell from producing acid. This may be written in symbolic form as:

$$\text{histamine} + \text{receptor} \longrightarrow \text{histamine-receptor} \longrightarrow \text{acid}$$

$$\text{cimetidine} + \text{receptor} \longrightarrow \text{cimetidine-receptor} \longrightarrow \text{no acid}$$

Two other peptic ulcer drugs have since been designed, both of which contain a five-membered ring with a long side-chain, that are stronger than cimetidine. They are known as ranitidine (Zantac™) which was produced in 1983 and famotidine (Pepcid™) produced in 1985. These newer drugs are four and twenty times respectively stronger than cimetidine in that the normal dosages are those factors smaller than that of cimetidine.

10.4 ASPIRIN

Many drugs have their origins in natural products. The willow tree (Latin name, *salix*), part of the family known as *Salicaceae*, is the origin of aspirin, acetylsalicylic acid, whose structure is shown in Fig. 10.5.

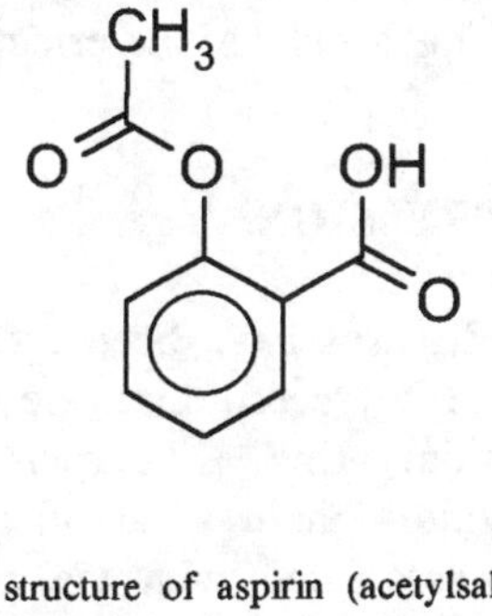

Fig. 10.5 The structure of aspirin (acetylsalicylic acid)

The pain killing properties of an extract of the bark of the willow tree have been known for a very long time. The pain killing agent is a compound of saligenin, $C_7H_8O_2$, and glucose, called salicin, which can be crystallized from an aqueous extract of willow bark. It was found that it was the saligenin which was the effective part of the compound and that the related compound, aspirin, was more gentle on the stomach than either saligenin or the the parent acid, salicylic acid, the structures of which are shown in Fig. 10.6.

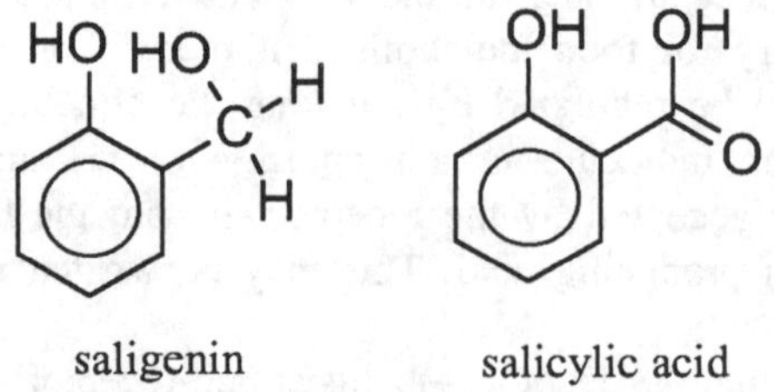

saligenin salicylic acid

Fig. 10.6 The structures of saligenin and salicylic acid

This is a classical example of how natural products with medicinal properties may give rise to even better products when modified chemically. The effective natural product is extracted and its structure identified. The effective part of the structure is determined by the testing of a variety of model compounds, the best model compound then undergoing modification to increase its effectiveness or to minimize any unwanted side effects.

In addition to its pain killing properties, aspirin is useful as an anti-inflammatory drug for patients suffering from rheumatism, gout and arthritis. Recently it has been shown that one 300 milligram aspirin tablet per day is an effective method of preventing secondary strokes. Strokes are associated with high blood pressure and **it is not recommended that the taking of aspirin should be adopted generally as a preventive measure.** The side effect of causing bleeding could possibly be associated with the causation of strokes if the treatment is prolonged. The pain killing effect of paracetamol is similar to that of aspirin and the drug is gentler on the stomach, but it does not have any of aspirin's other properties. It is dangerous in overdose, having a deleterious effect upon the liver.

10.5 TREATMENT OF DEPRESSION

Although diabetes and heart diseases have been afflicting the human race for many years, it is depressing to have to record that one modern affliction which is still on the increase is depression. This is one indication that in spite of the advantages of modern civilization there is an increase in unhappiness. There are some disadvantages that are associated with the present day way of life, just as there are side effects associated with most beneficial drugs. As in other branches of medicine, chemistry is useful in the production of antidepressant drugs. There is a range of effective drugs for the treatment of seriously affected patients with psychiatric illnesses such as schizophrenia. The most common form of depression is much less serious and is classified as anxiety, sufferers feeling anxious and possibly waking up too early. This mild form of depression is treated by the benzodiazepine family of drugs. The structure of the family of benzodiazepine molecules is shown in Fig. 10.7.

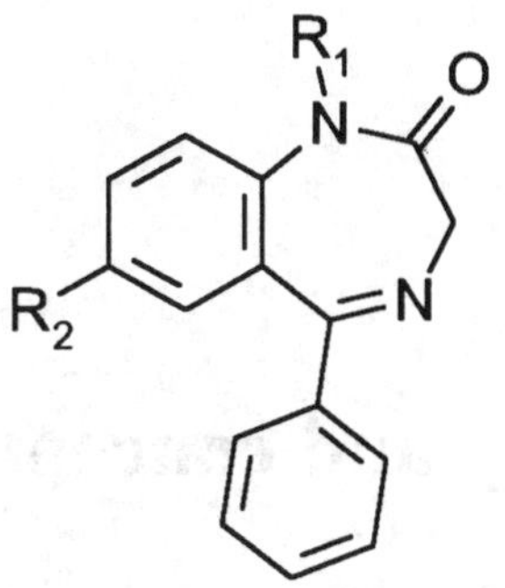

Fig. 10.7 The structure of benzodiazepine molecules

The groups R_1 and R_2 are varied to make the drugs Valium (R_1 = Cl, R_2 = CH_3) and nitrazepam (Mogadon, R_1 = NO_2, R_2 = H). Valium is an anti-anxiety drug prescribed for day-time use, whereas Mogadon is a mild but long-lasting hypnotic and is prescribed for patients who tend to wake up too early and find further sleep difficult to achieve. Although the benzodiazepine drugs are metabolized easily they do induce a dependence if taken over long periods.

10.6 IMPROVEMENTS IN HUMAN HEALTH

In addition to drug therapy, there are other ways in which the health and fitness of the human race may be improved. Continued efforts to purify the water which is drunk, i.e. a greater production and use of chlorine, and the air which is breathed will be beneficial. Legislation is important in the improvement in general health. Clean Air Acts have improved atmospheric conditions and the Health and Safety at Work Acts have forced companies to pay attention to dust control. Improvement in food hygiene should be encouraged. Accident prevention is important in the home and in the place of work. Individuals have the option to give up, or not to start, smoking. They have the options to drink less alcohol, eat a reasonable diet and to take part in physical activity. All these options have been shown to be desirable and can lead to the lengthening of lives which also have a better quality. A serious study of the mortality of scientists (*Lancet*, 1994, **343**, 296) has shown that, in spite of their everyday contacts with all manner of chemical substances, 72% reach the age of seventy, compared to non-scientists, only 67% of whom live to that age. Both groups had the same levels of alcohol intake, smoking and obesity. Individuals have the choice either to make chemistry work in their favour or take the consequences.

11

Chemistry: history and concepts

11.1 INTRODUCTION

This chapter consists of a brief outline of the history of chemistry and a concise description of current fundamental ideas upon which the modern subject of chemistry is based. It extends the ideas presented in Chapter 2 and consists of a qualitative non-mathematical treatment of the subject which is compatible with the contents of this book. It is written in six sections, the first containing a very short history of chemistry, the following five giving answers to the questions:

What are atoms and elements?
How are the elements classified?
What are molecules and compounds?
Why do chemical reactions occur?
How do chemical reactions occur?

Readers who find this chapter digestible are advised to re-read some of the earlier sections and thereby derive a fuller understanding of their contents.

11.2 HISTORICAL ASPECTS OF CHEMISTRY

The development of scientific theories. A history of elementary chemistry. Contributions of Empedocles and Aristotle. Transmutation of base metals into gold, the philosopher's stone and the elixir of life. Contributions of Paracelsus, Agricola, van Helmont and Boyle. The discovery of oxygen and the contributions of Priestley, Scheele and Lavoisier. Atomic theory from Democritus to Dalton. Dalton's law of multiple proportions. Atomic weights. The law of constant composition. Berzelius and the discovery of more elements. The contributions of Cavendish and Gay-Lussac. Gay-Lussac's law of combining volumes. Avogadro's hypothesis and the distinction between atoms and molecules. The mole. The chemical equation for the reaction between hydrogen and oxygen to give water. The Avogadro number. Relative molecular mass. A history of elemental combination.

Like all scientific theories, atomic theory has gone through many changes as a result of experimentation. Initial theories may be formulated from the results of experimental observations. Theories are then modified and improved by further experimentation. Better theories suggest more demanding experimental tests for themselves. By this cyclical process theories become established, but the process of refinement is never-ending.

This ultra-concise account of the history of atomic theory and the elements is included in this chapter, because it gives a general impression of how chemistry developed as a subject and the enormous human endeavour that has made it so important today.

The study of chemistry began with the discovery of fire, which allowed the chemical nature of foods to be altered by the process of cookery. Meat could be cooked by baking or roasting. In the Greek legend, Prometheus stole fire from heaven, a contribution commemorated in the name of element number 61; promethium, Pm. Zeus commanded Prometheus to make man. This he did, stealing fire from heaven to improve man's quality of life. Zeus did not approve of this action and had Prometheus bound in chains. The story is similar to that attributed to the Christian God who disapproved of Adam's gaining useful knowledge from the tree of life and banished him to the east of Eden (*Genesis*, iii, 22). In spite of such early set-backs, the progress of chemistry continued. The first cooks unknowingly carried out the early work on the denaturation of proteins, e.g. cooking meat, and the de-polymerization of carbohydrates, a process in which large molecules consisting of repeating units called monomers are broken down into smaller molecules, e.g. cooking potatoes, which converted their foods into much more tasty and edible forms. In the stone age, the baking of clay was noticed as the earth at sites of fires hardened. Clay-baking or pottery developed and allowed the invention and construction of vessels suitable for boiling water. The permanent dehydration of the clay brought about chemical changes which caused the resulting material to hold its shape. This added another variation of rendering food edible. The basic apparatus was now available for the development of cookery and experimental chemistry in general. The ancient chemists discovered the major processes of fermentation to yield wine and vinegar, and of metal extraction and also how to produce a large variety of organic substances including waxes, oils, soap, perfumes and medicines.

There are no records of any early attempts to rationalize the findings of early chemistry, but modern atomic theory has its roots in ancient Greek thoughts. The early thinkers suffered from a tendency to ignore experimental results and had an aversion to experimentation. This hindered the proper development of their ideas. Although various philosophers had suggested that the basis of the material world was one or other of the 'primary bodies' of air, earth, fire and water (these are discussed in Chapters 4, 5, 6 and 7 respectively), it was Empedocles who concluded that all four were of equal importance. He lived between 492-432 B.C. in the city now known as Agrigento in Sicily. Although he despised experimentation, he observed the properties of the klepshydra, a pipette-like tube used for the extraction of wine or water from jars. When a finger was placed over one end of a tube, he noticed that water did not enter the tube when the other end was immersed in the liquid. Water did enter the tube if the finger was then removed from its end. He was able to conclude that air had body and was a material substance. This important advance in the understanding of matter arose from a simple experimental observation.

Aristotle (384 - 322 B.C.) re-named the primary bodies as **elements** and developed the four element theory. The theory was to last until the end of the eighteenth century A.D. For a theory to persist for so long without contradiction or modification indicates that it must have been able to explain many phenomena to the satisfaction of observers. At the end of the eighteenth century, alchemists were still very active and experimental results were very numerous. Respected journals were not available, so the dissemination of experimental information and its discussion was severely limited. The success of the four-element theory lay in the general association of air, water and earth with the three states of matter; gas, liquid and solid. Fire was observed during some of the conversions of solid substances into water and 'air', and caused other conversions to occur. Wood burns with a flame and is converted to 'air' (carbon dioxide), water and 'earth' (ash). Empedocles shared the same fate when he jumped into the crater of Mount Etna. He thought of himself as a god who could withstand fire and, therefore, could survive such an act. His theory was conceptually flawed. Theories in which gods participate cannot be regarded as scientific, because they can never be tested satisfactorily. Gods have the man-made characteristic of being all things to all men. The result of any test of a theory of which they form a part could be manipulated so as to be consistent with the theory-holder's belief. The experiment that Empedocles carried out upon himself did not increase our information about gods. Gods may or may not survive very high temperatures, but there is no doubt at all about human frailty under such conditions.

The alchemists' activities were dominated by two great human aspirations. One was to be rich, the other to be healthy. The first gave rise to the urge to transmute or convert base metals into gold. Much alchemical effort was expended in the search for the **philosopher's stone** which would bring about the required transmutations. The alchemists were encouraged by the success of experiments in which silver was extracted from galena and gold from iron pyrites. Galena is the mineral, lead sulfide (PbS), some deposits containing low levels of silver. Likewise, gold is found in some deposits of iron pyrites (FeS_2), the mineral known commonly as fool's gold, which is a compound containing iron and sulfur. They were convinced that they could transmute iron into copper by adding metallic iron to a solution of blue-stone, i.e. copper sulfate ($CuSO_4$). In such an experiment, the iron dissolved to give a solution of iron sulfate, the metallic copper being displaced from the original solution of copper sulfate. An equation representing this reaction may be written in words as:

$$\text{iron} + \text{copper sulfate} \longrightarrow \text{copper} + \text{iron sulfate}$$

or as a properly balanced chemical equation:

$$Fe + CuSO_4 \longrightarrow Cu + FeSO_4$$

or even as an 'ionic' equation, omitting the sulfate ion (SO_4^{2-}) since it takes no part in the transfer of two electrons from the iron to the copper(II) ion:

$$Fe + Cu^{2+} \longrightarrow Cu + Fe^{2+}$$

This reaction is used currently by copper producers and is discussed in Chapter 5.

The second major alchemical activity was the search for the **elixir of life**, the 'big medicine' which would solve all our bodily problems including that of the degeneration of tissue and organs that is known as aging. Much experimental work was carried out on the medical properties of substances. The scientific development of the two aspirations, in modern terms, would be called inorganic and organic chemistry respectively. It was the German medical man, Theophrastus Bombast von Hohenheim, known as Paracelsus (1493-1541), who concluded that animal and vegetable bodies were constituted from the three elemental components, air, water and earth. He regarded the health of an organism to be dependent upon the presence of proper proportions of the three elements. If the ratios were not ideal the organism would be diseased. It was Paracelsus who brought together the disparate activities of the inorganic and organic chemists and helped to form the basis of the subject of chemistry.

In the mid-fourteenth century only thirteen elements, in the present day use of that term, were known. They were the two non-metals, carbon and sulfur, and eleven metals, copper, silver, gold, iron, tin, antimony, mercury, lead, arsenic, zinc and bismuth. Arsenic is now regarded as a metalloid, its character being on the border between those of metals and non-metals. A vast amount of chemical knowledge was being assembled and classified. Georg Bauer (1494-1555), known as Georgius Agricola, published his *De Re Metallica* in 1556. This was the first book on mining and metallurgy, the extraction of metals from their ores, and the first based on the proper recording of experimental observation which would now be regarded as the scientific approach.

To challenge a theory which had been in existence for nineteen centuries required someone with considerable intellect. Such was Johann Babtist van Helmont (1587-1657). He discounted fire as an element, but retained the view that air and water were elements. He was the first to distinguish between the different kinds of 'air', which he called gases. The word gas was derived from the Greek word meaning chaos, the connexion being very helpful to our understanding of the gaseous state; a chaotic movement of molecules. He recognized that the gas given off in fermentation, carbon dioxide, was different from common atmospheric air. Even so, alchemical ideas lingered on for some time until the publication in 1661, of *The Skeptical Chymist* by Robert Boyle (1627-1691). Boyle put forward the view that to be an element, a body must not be capable of further sub-division. From a combined body, i.e. a compound, elements could be obtained from which the combined body could be regenerated. He was enunciating the current views that 'an element is a substance that cannot be divided up to give simpler substances' and that 'a compound is a substance containing more than one element'. Boyle taught that the study of chemistry was an essential part of the greater study of *Nature* (not the scientific journal!) and that the application of scientific method was the proper route to the advancement of knowledge. At that time the elements identified were still thirteen in number. A too general theory, such as the four element theory makes progress difficult and slow. The theory was too simple and too comprehensive to be found wanting by most chemical experiments being carried out at the time.

After Boyle had set the scientific community on the right track, there was a burgeoning of chemical experimentation. The nature of air remained a mystery until 1774, when Joseph Priestley (1733-1804) discovered oxygen. In 1772, Priestley had already established that burning a candle caused a deterioration in the quality of the air and reduced its volume. He found that mice caused a deterioration in the quality of the air they breathed. He also demonstrated that the quality of the air, reduced by either a burning candle or by the breathing of mice, could be restored by the action of green plants. This was the first observation of photosynthesis. Priestley made oxygen by decomposing 'red precipitate' (mercuric oxide, HgO), heating it with the sun's rays which he concentrated by using a magnifying glass. In 1781 he showed that water resulted from the burning of hydrogen in air but he was unable to explain the reaction. His thoughts were retarded by the theories of the day.

Henry Cavendish (1731-1810) discovered hydrogen in 1766 and Daniel Rutherford discovered nitrogen in 1772. By 1784, Cavendish had shown that one volume of oxygen required exactly two volumes of hydrogen for complete reaction when burned and that the product was pure water. This experiment and its results were of great significance for the development of atomic and molecular theory. At the same time as Priestley and Cavendish were carrying out their researches, Karl Wilhelm Scheele (1742-1786) in Sweden, and Antoine Laurent Lavoisier (1743-1794) in France, were doing similarly distinguished work. Priestley, Scheele and Lavoisier should all be credited with the discovery of oxygen, all three chemists' work contributing fundamentally to the basis of the modern subject. Priestley and Scheele were honoured in their own countries, but Lavoisier was beheaded during the French revolution. Lavoisier's work is now recognized, even by the French, as being one of the most significant contributions to modern chemistry. His book, *The Elements of Chemistry*, which was published in 1789, was the first proper textbook on the subject.

Lavoisier made mercuric oxide by gently heating mercury in contact with a known volume of air. He found that, in producing mercuric oxide, about one fifth (air contains 20.8% by volume of dioxygen molecules, O_2) of the air was used up and that the residual gas would not support life, nor would it allow anything to burn in it. He then separated the mercuric oxide from the residual metal and heated it strongly. The mercuric oxide decomposed into metallic mercury and a gas which was of a volume equal to the loss of volume by the air used in the first experiment. He named the gas oxygène (from the Greek *oxy genes* = acid forming, some oxides when dissolved in water giving an acidic solution, e.g. sulfur trioxide, SO_3, which gives sulfuric acid, H_2SO_4). Further work led Lavoisier to his theory of combustion, which was essentially that a metal was converted to its oxide when it burns in air. This led to the proper interpretation of the experimentally observed increase in weight when a metal burns. The oxide of a metal is heavier than its metal content by the amount of oxygen required to produce the oxide. The previously held theory, which

Lavoisier's work destroyed, was that a metal became heavier when burnt because of the emission of a fire-like substance known as **phlogiston**. To be consistent with the experimental observations, it was necessary for phlogiston to have a negative weight. The realization that the phlogiston theory was not viable allowed great progress to be made in the understanding of chemical reactions.

By 1774, only eight more elements had been added to the thirteen known by 1500. These were phosphorus, platinum, cobalt, nickel, magnesium, hydrogen, nitrogen and oxygen. What was required to allow chemistry to expand, was a reasonable explanation of all the knowledge that had been accumulated since Boyle initiated the scientific method of investigation. An atomic theory was needed that explained the difference between elements and compounds and the ratios in which elements combined to form compounds. It is necessary to go back to the early Greek philosophers to find the origins of atomic theory and to the ideas expressed by Democritus in particular.

Democritus (~460-370 B.C.) is generally credited with being the main originator of early Greek atomic theory which proposed that all matter had an atomic basis. A substance was composed of specific atoms, characteristic of that substance and separated by spaces. It was impossible to sub-divide the substance beyond its individual atoms. The 'atoms' conceived by Democritus were equivalent to our present day concept of molecules.

The transformation of atomic ideas into a quantitative theory able to explain the nature of elements and compounds was announced in 1803 by John Dalton (1766-1844) in a lecture at the Royal Institution in London. He had already published his first table of relative atomic weights in a paper given at the Manchester Literary and Philosophical Society earlier that year. He found that atoms of different elements have different relative weights. The currently used term, relative atomic mass or RAM, is the mass of the element compared to that of the main isotope of carbon which is taken to be exactly 12. This is explained more fully later in this chapter. He also stated that *'the relative atomic weights of the elements are the proportions by weight in which these elements combine, or some simple whole-number multiple or sub-multiple of these'*. This was a statement of the **law of multiple proportions.** Dalton proposed symbols to represent the elements and, although his symbols were soon replaced by the ones now in use, were a major advance in the representation of chemical reactions. It became possible (using our present day symbols) to explain the atomic ratios that exist in compounds such as nitric oxide, NO (one atom of nitrogen and one atom of oxygen), and nitrous oxide, N_2O (two atoms of nitrogen and one atom of oxygen). Using their present-day atomic weights (N = 14, O = 16), it became possible to understand why nitric oxide, for example, contained nitrogen and oxygen in the ratio of 7 to 8, i.e. 14 to 16, by weight. Likewise, in nitrous oxide the nitrogen-to-oxygen ratio was 7 to 4, i.e. (2 x 14) to 16 or, more simply, 7 to 4.

The **law of constant composition**, for any given compound, sometimes referred to as the **law of definite proportions**, was immediately understandable. Much of the basic work on the law of constant composition was carried out by Carl Friedrich Wenzel (1740-1793), a chemist at the Meissen porcelain factory. The law of multiple proportions, for any two or more combining elements as explained above for the nitrogen oxides, was also understandable. The nitrogen contents of nitrous oxide

(N_2O) and nitric oxide (NO) exhibited a 2:1 simple integral atomic ratio for a given oxygen content, as expected from the two formulae.

The firm establishment of both these fundamental laws of chemical combination was carried out by the great Swedish chemist, Jöns Jacob Berzelius (1779-1848), whose work was carried out more accurately than was Dalton's. Berzelius also introduced the modern nomenclature for the symbols of the elements that is still in use; the initial letter and usually one other letter from the element's name providing the symbol. By 1818, when Berzelius published his theories of chemical proportions, fifty two elements had been identified of the eventual total of eighty eight that are now known to occur naturally.

The molecular nature of elements was not clear at the time of Dalton's atomic theory. It was not known that the oxygen and nitrogen in the air, although they were elements, consisted of the diatomic molecules, O_2 and N_2, now known as dioxygen and dinitrogen respectively. Further understanding came from the work of Gay-Lussac and Alexander von Humbolt in 1805, who found that one volume of oxygen reacted with exactly two volumes of hydrogen to form water, confirming the observations of Cavendish. Gay-Lussac was able to announce a general law in 1808, that gases combine in simple ratios by volume and give gaseous products whose volumes bear a simple relation to the volumes of any gaseous reactants. The explanation of the Gay-Lussac law of combining volumes was given in 1811 by Amadeo Avogadro. Avogadro's celebrated hypothesis supposed that '*equal volumes of gases, at the same temperature and pressure, contained equal numbers of molecules*'; i.e. groups of atoms bonded together. The recognition that atoms and molecules were different, cleared up much of the misunderstanding about chemical reactions.

By the early nineteenth century, the concepts of atoms and molecules, elements and compounds, were generally accepted as those now currently held. One major concept which developed from all the work carried out in the early part of the nineteenth century was the 'mole', discussed in Chapter 2. It had become obvious that chemical reactions between elements and/or compounds occurred in such a way that the ratios of the combining substances, whether they were solids, liquids or gases, were simply related to their relative atomic or molecular masses. In addition, if the reactants were gaseous, their combining volumes bore a simple relation to each other. The extension of Avogadro's idea to the concept of the mole, applicable to gases, liquids and solids, clarified the whole of reaction chemistry. The mole represents the amount of a substance which, measured in grams, equals the relative molecular mass of that substance. Thus, one mole of carbon is twelve grams of the element and one mole of oxygen is thirty two grams of the diatomic gas. The O_2 molecule, strictly known as dioxygen, has a relative molecular mass (RMM) which is twice the relative atomic mass (RAM) of atomic oxygen. The strict definition of the mole is that it is **the amount of substance which contains as many elementary entities as there are atoms in 0.012 kilograms of the mass-12 isotope of carbon**. In this definition, it is important to appreciate that 'elementary entities' may be free atoms of an element, molecules of an element or compound, ions with positive or negative charges, electrons, other particles or specified groups of such particles.

The earliest reference to the combination of atoms is contained in the work of Lucretius (~100 - 55 B.C.) who wrote a very long poem, amounting to 7400 lines, about the physical theories of Epicurus (342 - 270 B.C.). The poem was the main source of transmission of the early Greek theories to the readers of the Middle Ages. Epicurus was mainly concerned with the attainment of a happy life, but had time between his rather austere meals to develop the ideas of Democritus considerably. It is difficult to identify the contributions of Lucretius to the Epicurean theories. His poem was written around 57 B.C. It contained the first ideas of the first law of thermodynamics which is the law of conservation of energy and matter, dealt with later in this chapter. He reported that atoms differ in size shape and weight. He came to the conclusion that there must be an upper limit to the size of an atom, reasoning that if this were not so, some atoms would be large enough to be visible. Atoms which came into contact with each other could become attached and further accretions would form 'things'. There was a rudimentary attempt to explain how atoms could interact to form molecules. Atoms possessed hooks which allowed them to combine. This was the first theory of the combining ability of atoms, this being known as **valency**. It is still used in explaining bonding and valency to junior school pupils. There was little development in the theory until 1916 when Gilbert Newton Lewis postulated the **stable octet theory**. This was that molecules achieved stability when the central atom possessed a share in an octet of electrons. He also defined the covalent bond, as it is now known, as the sharing of two electrons between two atoms. The idea was very close to that held today. Lewis's ideas were enlarged upon by Langmuir in 1919, who coined the term covalent bond, and by Sidgwick in 1927. A major advance was made between 1931-1933 by Linus Pauling, who has received two Nobel Prizes. One was the chemistry prize awarded in 1954, somewhat belatedly, for his work on chemical bonding, the other was the peace prize awarded in 1962. He has stated that 'every aspect of the world today, even politics and international relations, is affected by chemistry'. It is hoped that readers of this book will come to agree with him. One other name stands out and should be mentioned as his work forms the current basis of almost all the research being carried out into the very abstruse and sophisticated theory of chemical bonding; Robert S. Mulliken, whose work was carried out in 1933. He was awarded the Nobel prize for chemistry in 1966, an even longer wait than Pauling had endured. The highly mathematical nature of current theory cannot be dealt with in this book, but the descriptions of bonding are based very largely upon Mulliken's theory enhanced by the power that modern computers offer in the solutions of the complex equations.

11.3 ATOMS AND ELEMENTS

Atoms, elements, molecules and compounds. Atomic numbers, relative atomic masses and isotopes. Names, atomic numbers, relative atomic masses and the dates of discovery or synthesis of the elements.

As indicated in Chapter 2, all material substances consist of atoms. An element consists of atoms characteristic of that element and no other. An atom of an element is defined as the smallest particle which possesses the characteristics of that element. Each element has its own symbol, with a subscript (for molecular species) indicating the number of atoms of that element contained by the molecule. Some elements exist in the form of single atoms, e.g. the gases helium and neon, while others exist as molecules which consist of two or more atoms linked together. For example, oxygen is normally the gaseous diatomic molecule, O_2 (dioxygen is its formal name), and solid sulfur consists of the cyclic molecule, S_8 (octasulfur), which contains eight sulfur atoms joined together in the form of a puckered ring as shown in Fig. 11.1.

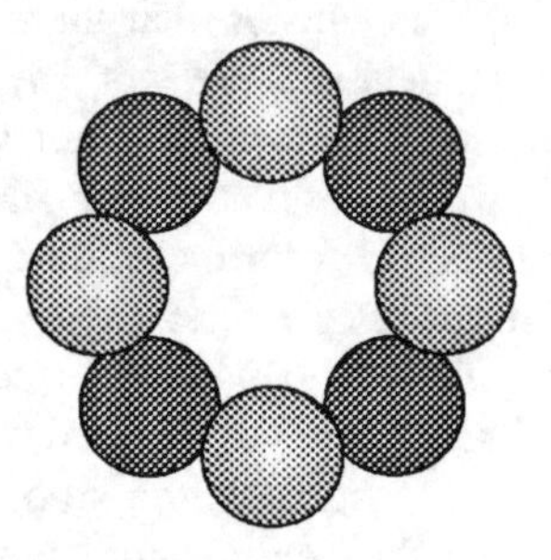

Fig. 11.1 The structure of octasulfur, S_8; the two sets of differently shaded atoms occupy parallel planes

Metallic elements exist as solid crystalline arrays consisting of a very large number of atoms arranged in an orderly fashion, most metal structures being similar to the arrangement of oranges on the fruit stall. Since there are no identifiable molecules in such metallic structures, metallic elements are usually symbolized by the subscript-less atomic symbol, e.g. copper metal is symbolized as Cu.

Some chemical compounds consist of molecules which may contain two or more different atoms. Others consist of infinite arrays of the constituent atoms in which individual molecules cannot be identified. A molecule is the smallest particle of a compound which can exist. The molecule of methane, which is the main constituent of natural gas, has the formula, CH_4, indicating that it contains one atom of carbon and four atoms of hydrogen. Molecules have specific shapes, that of methane being tetrahedral as shown in Fig. 2.4. The four hydrogen atoms are at the vertices of a regular tetrahedron with the carbon atom at its centre. An example of a compound, although having an apparently simple formula, which does not exist as individual molecules, but is an infinite array of the constituent atoms is silicon carbide (carborundum), SiC. The solid contains equal numbers of silicon and carbon atoms arranged regularly throughout its structure, each atom being surrounded tetrahedrally by four atoms of the other kind.

Eighty eight elements are known to occur naturally. Other elements do exist, but these are synthetic and have short lifetimes because they are intensely radioactive and undergo radio-active decay. Radioactivity is described in Section 3.9. The cumulative total of known elements varied with time in a very irregular manner, as can be seen from the graph in Fig. 11.2.

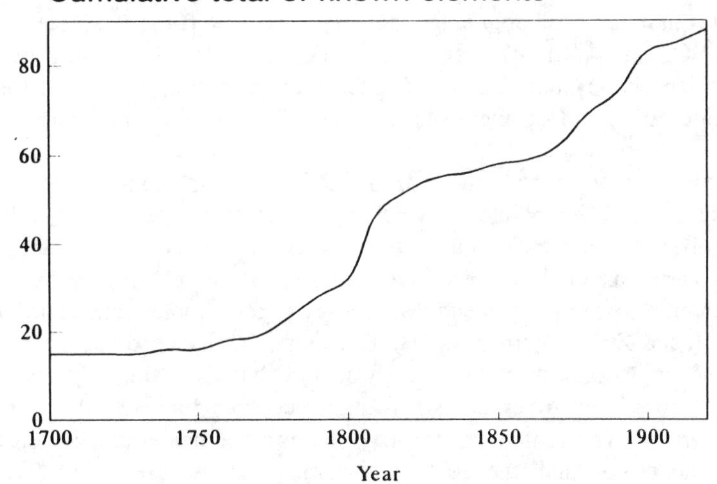

Fig. 11.2 The cumulative total of known elements as it has varied with time

Apart from the eleven elements known from antiquity (carbon, sulfur, copper, silver, gold, iron, tin, antimony, mercury, lead and zinc) and the discovery of arsenic, bismuth, phosphorus and platinum in the years from 1500 to 1700, the increase in the number of known elements started in 1735 with the discovery of cobalt. Nickel was discovered in 1751, followed by magnesium in 1755, hydrogen in 1766 and nitrogen in 1772. After the discovery of oxygen, the rate of discovery of new elements increased. A further and very significant increase occurred after Dalton's atomic theory was put forward in 1803 and gained acceptance. The rate of discovery slowed down after 1810, the main reason for this being the lack of analytical methods and the low abundances of the yet-to-be-discovered elements. As analytical methods improved, in the time after 1860, the remainder of the elements were discovered, the last stable one being rhenium which was found in 1925. New elements *discovered* after 1925 have all been synthetically produced by nuclear reactions.

The two important characteristics of an element are its atomic number, Z, each element having a different value of Z, and its relative atomic mass, RAM, originally known as atomic weight. It is current practice to explain the structure of atoms in terms of a positively charged nucleus which is surrounded by a sufficient number of negatively charged electrons to make the atom neutral. The nucleus consists of a number of protons, each with a positive charge of +1, represented by the value of the atomic number, Z, and a number of neutrons, which are neutral particles, equal to the value of $A - Z$, where A represents the mass number which is the nearest whole number to the exact relative atomic mass of the atom. Protons and neutrons both have masses which are very close to unity, i.e. 1, on the relative atomic mass scale. For example, the element fluorine (F) has an atomic number of 9 and a mass number of 19. The exact relative atomic mass of the fluorine atom is 18.9984, accurate to the fourth decimal place. The fluorine nucleus contains 9 protons and $19 - 9 = 10$ neutrons.

The mass number, A, is quoted as a left-hand superscript to the element's symbol, the atomic number being indicated as a left-hand subscript as in the examples, ${}^{35}_{17}Cl$ and ${}^{37}_{17}Cl$. The atom is rendered neutral by Z negatively charged electrons which occupy the volume of space around the nucleus, e.g. 17 electrons for each chlorine atom. Most elements consist of a mixture of naturally occurring isotopes.

The existence of neutrons made possible the understanding of isotopy. Isotopes of an element are different atoms which have the same value of Z, otherwise they would be different elements, but have different numbers of neutrons so that the value of A varies as in the case of the two isotopes of chlorine quoted above. The simplest example of the phenomenon of isotopy is the simplest element, hydrogen, for which $A = 1$ and $Z = 1$, written as ${}^{1}_{1}H$. The most abundant isotope consists of nuclei consisting of single protons with no neutrons so that the value of A is 1. That kind of hydrogen atom, sometimes called protium, accounts for 99.985% of the naturally occurring element. The remainder consists of atoms of deuterium, which have nuclei containing one proton and one neutron, written as ${}^{2}_{1}H$ and sometimes as the special symbol D. The value of A for deuterium is two. There is a third isotope of hydrogen whose nucleus contains two neutrons so that its value of A is three, written as ${}^{3}_{1}H$ or T. This isotope is called tritium but only exists naturally to the very small extent of about one atom in 10^{18}. The nuclear compositions of the three isotopes of hydrogen are given in Table 11.1.

Table 11.1 - The composition of the nuclei of the three isotopes of hydrogen

name of isotope	atomic number (value of Z)	mass number (value of A)	number of neutrons = value of $A - Z$
protium	1	1	0
deuterium	1	2	1
tritium	1	3	2

Atomic weights or relative atomic masses, as they are now properly known, are given in Tables 11.2 - 11.6 for the first 103 elements. They not absolute weights, or masses, of individual atoms. They are relative masses of atoms compared to the mass of the mass-12 isotope of carbon, which is taken to have a mass of exactly 12. The carbon scale was chosen because of the existence of a range of suitable gaseous carbon compounds which can be used to calibrate mass spectrometers, the instruments now used almost exclusively for the determination of relative atomic and molecular masses. Relative atomic masses can be determined to an accuracy where at least the fourth decimal place is significant. There are applications for very accurately determined relative atomic masses and they are used in this book when necessary. Most elements are a mixture of naturally occurring isotopes, the relative contributions of each isotope being independent of the source of the element in most cases. The chemical relative atomic mass of an element, as quoted in Tables 11.2 -

11.6, is the weighted average of the values for the natural mixture. For example, the element chlorine has an atomic number of 17 and has two naturally occurring isotopes. One isotope, which is present to the extent of 75.77%, has a relative atomic mass of 34.968852, for which A = 35. The other isotope has a relative atomic mass of 36.965903, A = 37. The chemical relative atomic mass is therefore calculated as:

$$\text{RAM (Cl)} = (0.7577 \times 34.968852) + (0.2423 \times 36.965903)$$

$$= 35.452737$$

The relative atomic masses of individual isotopes of an element are always very close to being a whole number, the two isotopes of chlorine being examples. The two isotopes have exact relative atomic masses which are very close to being the whole numbers 35 and 37 respectively. The reason for the RAM's closeness to whole numbers is because the constituents of the nuclei of atoms, protons and neutrons, have exact relative masses which are very close to being 1 on the relative atomic mass scale. The presently accepted relative atomic masses of the hydrogen atom, which consists of one proton and one electron, and the neutron are 1.007825 and 1.008665 respectively. The relative mass of an electron is only 1/1837 of that of the hydrogen atom.

The known elements, both natural and synthetic, at the present time, in the order of their atomic numbers, are given in Tables 11.2 - 11.6, together with their relative atomic masses (RAM) and the dates of their discovery or synthesis, in the cases of elements which are prepared by nuclear reactions rather than those which exist naturally. The significance of the division of the elements into the tables and of the divisions of the elements in the tables into sub-groups is explained in detail below. Table 11.2 contains information about the first eighteen elements and is subdivided into three sets of elements; (i) hydrogen and helium, (ii) the eight elements from lithium to neon and (iii) the eight elements from sodium to argon. Within each set of eight elements the individual members are chemically very different from each other. The respective members of the two sets of eight elements, however, exhibit close chemical similarities to each other. For example, lithium and sodium are both soft metals which react readily with water and neon and argon are both gases with no chemical reactivity.

Table 11.2 - The first eighteen elements in order of their atomic numbers, their symbols, relative atomic masses (RAM, three significant decimal places) and dates of discovery if known

element name and symbol	atomic number	RAM	date of discovery
hydrogen, H	1	1.008	1766
helium, He	2	4.003	1895
lithium, Li	3	6.941	1817
beryllium, Be	4	9.012	1797
boron, B	5	10.811	1808
carbon, C	6	12.011	*
nitrogen, N	7	14.007	1772
oxygen, O	8	15.999	1774
fluorine, F	9	18.998	1886
neon, Ne	10	20.18	1898
sodium, Na	11	22.99	1807
magnesium, Mg	12	24.305	1755
aluminium, Al	13	26.982	1825
silicon, Si	14	28.086	1824
phosphorus, P	15	30.974	1669
sulfur, S	16	32.006	*
chlorine, Cl	17	35.453	1774
argon, Ar	18	39.948	1894

* these elements were known before records were kept

Hydrogen and helium, the elements from boron to neon, and those from phosphorus to argon, are non-metals. Silicon is classified as a metalloid. Its properties are largely those exhibited by a non-metal, but the element has important semi-conducting properties, critical to the manufacture of electronic chips used in computers. The other elements, lithium and beryllium, and sodium, magnesium and aluminium, are metals.

Table 11.3 - Elements 19-36 in order of their atomic numbers, their symbols, relative atomic masses (RAM, three significant decimal places) and dates of discovery if known

element name and symbol	atomic number	RAM	date of discovery
potassium, K	19	39.098	1807
calcium, Ca	20	40.078	1808
scandium, Sc	21	44.956	1879
titanium, Ti	22	47.88	1791
vanadium, V	23	50.942	1801
chromium, Cr	24	51.966	1780
manganese, Mn	25	54.938	1774
iron, Fe	26	55.847	~2500 BC
cobalt, Co	27	58.933	1735
nickel, Ni	28	58.69	1751
copper, Cu	29	63.546	~5000 BC
zinc, Zn	30	65.39	*
gallium, Ga	31	69.723	1875
germanium, Ge	32	72.61	1896
arsenic, As	33	74.922	1250
selenium, Se	34	78.96	1817
bromine, Br	35	79.904	1826
krypton, Kr	36	83.80	1898

* zinc was identified as an element before 1500, but was known as a constituent of brass (a copper-zinc alloy) around 20 BC

The elements in Table 11.3 are metals, with the exception of germanium, arsenic, and selenium, which are metalloids, and bromine and krypton, which are non-metals.

Table 11.4 - Elements 37-54 in order of their atomic numbers, their symbols, relative atomic masses (RAM, three significant decimal places) and dates of discovery

element name and symbol	atomic number	RAM	date of discovery
rubidium, Rb	37	85.478	1861
strontium, Sr	38	87.62	1808
yttrium, Y	39	89.906	1794
zirconium, Zr	40	91.224	1789
niobium, Nb	41	92.906	1801
molybdenum, Mo	42	95.94	1781
technetium, Tc	43	98.906	1937*
ruthenium, Ru	44	101.07	1808
rhodium, Rh	45	102.906	1803
palladium, Pd	46	106.42	1803
silver, Ag	47	107.868	~3000 BC
cadmium, Cd	48	112.411	1817
indium, In	49	114.82	1863
tin, Sn	50	118.71	~2100 BC
antimony, Sb	51	121.75	~1600 BC
tellurium, Te	52	127.60	1783
iodine, I	53	126.904	1811
xenon, Xe	54	131.29	1898

* technetium (Greek technikos = artificial) does exist in nature in very minute amounts as a fission product of natural uranium, but was first isolated from a sample of molybdenum which had been bombarded with accelerated deuterium nuclei. Its synthesis is expensive, but considerably cheaper than its extraction.

The elements of Tables 11.3 and 11.4 are subdivided into sets containing 2, 10 and 6 members, the sets of 10 elements coming between the eight elements which have close chemical similarities respectively to the two sets of eight elements in Table 11.2. The respective members of the two sets of ten elements exhibit close chemical similarities to each other. For example, the chemistry of of chromium is similar to that of molybdenum, both being the fourth member of their respective sets of ten elements. The sets containing ten elements are known as transition elements. The elements in Table 11.4 are all metals, except for antimony and tellurium, which are classified as metalloids, and iodine and xenon which are non-metals.

Table 11.5 - Elements 55-86 in order of their atomic numbers, their symbols, relative atomic masses (RAM, three significant decimal places) and dates of discovery

element name and symbol	atomic number	RAM	date of discovery
caesium, Cs	55	132.905	1860
barium, Ba	56	137.327	1808
lanthanum, La	57	138.906	1839
cerium, Ce	58	140.115	1803
praseodymium, Pr	59	140.908	1885
neodymium, Nd	60	144.24	1885
promethium, Pm	61	146.915	1945*
samarium, Sm	62	150.36	1879
europium, Eu	63	151.965	1901
gadolinium, Gd	64	157.25	1880
terbium, Tb	65	158.925	1843
dysprosium, Dy	66	162.50	1886
holmium, Ho	67	164.930	1878
erbium, Er	68	167.26	1842
thulium, Tm	69	168.934	1879
ytterbium, Yb	70	173.04	1878
lutetium, Lu	71	174.967	1907
hafnium, Hf	72	178.49	1923
tantalum, Ta	73	180.948	1802
tungsten, W	74	183.85	1783
rhenium, Re	75	186.207	1925
osmium, Os	76	190.2	1803
iridium, Ir	77	192.22	1803
platinum, Pt	78	195.08	pre-1700
gold, Au	79	196.967	~3000 BC
mercury, Hg	80	200.59	~1500 BC
thallium, Tl	81	204.383	1861
lead, Pb	82	207.2	~1000 BC
bismuth, Bi	83	208.980	~1500
polonium, Po	84	208.982	1898
astatine, At	85	209.987	1940*
radon, Rn	86	222.018	1900

*promethium and astatine are synthetically produced elements and although a very slight trace of promethium exists in uranium ores as a fission product of natural uranium, its synthesis is far cheaper than its extraction.

The subdivision of Table 11.5 into sets of 2, 14, 10 and 6 elements indicates an extra complexity in the periodicity of the chemical nature of the elements. The set of 14 elements represents the lanthanides, elements that are almost chemically identical to lanthanum. The lanthanides, and the set of 10 transition elements from lutetium to mercury, separate the eight elements which show chemical similarities to the respective members of the other sets of eight in the previous Tables 11.2, 11.3 and 11.4. The elements in Table 11.5 are all metallic, except for astatine and radon, which are non-metals.

Table 11.6 - Elements 87-103 in order of their atomic numbers, their symbols, relative atomic masses (RAM, three significant decimal places) and dates of discovery or synthesis

element name and symbol	atomic number	RAM	date of discovery
francium, Fr	87	223.02	1939 synth
radium, Ra	88	226.025	1898
actinium, Ac	89	227.028	1899
thorium, Th	90	232.038	1815
protactinium, Pa	91	231.036	1917
uranium, U	92	238.029	1789
neptunium, Np	93	237.048	1940 synth
plutonium, Pu	94	244.064	1940 synth
americium, Am	95	243.061	1944 synth
curium, Cm	96	247.070	1944 synth
berkelium, Bk	97	247.070	1949 synth
californium, Cf	98	251.08	1950 synth
einsteinium, Es	99	252.083	1952 synth
fermium, Fm	100	257.095	1952 synth
mendelevium, Md	101	258.099	1955 synth
nobelium, No	102	259.101	1958 synth
lawrencium, Lr	103	260.105	1961 synth

Table 11.6 has subdivisions of 2, 14 and 1 elements, the set of 14 elements being known as actinides. All the elements in Table 11.6 are metals. Elements with higher atomic numbers than lawrencium (104 to 109) have been synthesized, are intensely radioactive, and are of theoretical significance only at the present time. They are all metals. The symbols of the elements are those suggested by Berzelius, for the ones known in his time, with a few alterations. The elements discovered since have symbols related to their names which are usually obvious except when the Latin name is used as a basis for the symbol. The Latin names of sodium (*natrium*), potassium (*kalium*), iron (*ferrum*), copper (*cuprum*), silver (*argentum*), tin (*stannum*), antimony

(*stibium*), gold (*aurum*), mercury (*hydrargyrum* = liquid silver) and lead (*plumbum*) are used as the basis for their symbols. The only other element name and symbol which appear to be unconnected are tungsten and W, the symbol being derived from the German name for the element, *wolfram*. Of the transuranic elements, i.e. those with values of Z greater than 92 which is the value for uranium, neptunium and plutonium are named after the planets Neptune and Pluto. Those from americium to lawrencium were synthesized in the Lawrence laboratory of the University of California at Berkeley and were named after the continent (i.e. America), state, university and laboratory of their births, with the scientists, Curie, Einstein, Fermi, Mendeleev and Nobel (inventor of dynamite), also being commemorated. Edward Lawrence (1939 Nobel prize for physics) invented the cyclotron; the machine which was used for element synthesis. The commemorative tradition has been extended to elements 104 (Dubnium, Db), 105 (Joliotium, Jl), 106 (Rutherfordium, Rf), 107 (Bohrium, Bh), 108 (Hahnium, Hn) and 109 (Meitnerium, Mt). Element 104 was synthesized at the Joint Nuclear Research Institute at Dubna in Russia. Frederic and Irène Joliot-Curie shared the 1935 Nobel prize for chemistry for their work on synthetic radioactive isotopes. Ernest Rutherford's experiments provided the foundation for our current understanding of atomic structure; he was awarded the 1908 Nobel prize for chemistry. Niels Bohr (1922 Nobel prize for physics) made major contributions to early atomic theory. Otto Hahn and Lise Meitner shared in the discovery of nuclear fission, as did Strassman, Hahn being the recipient of the 1944 Nobel prize for chemistry. The names of elements 104-109 have yet to be ratified by the IUPAC.

11.4 ELEMENTS AND THE PERIODIC CLASSIFICATION

The periodic classification of the elements. Valency. The metallic/non-metallic character of elements and their physical states under standard conditions.

Although atomic theory was accepted by the early part of the nineteenth century and the ideas relating to elements and compounds had been clarified, there was a lack of general understanding of the structure of the atom at that time and consequently the variations in the nature and properties of the known elements and their compounds were not well understood. The classification of the elements in terms of their properties culminated, in 1869, in the publication of the Periodic Law by Dimitri Ivanovitsch Mendeleev (1834-1907) who worked in St.Petersburg. The Law was stated as '*the properties of the elements are in periodic dependence upon their atomic weights*'. An almost identical generalization was published independently by Lothar Meyer in 1870. It was observed that there was a periodicity in physical and chemical properties. When the importance of the atomic number was realized, it became necessary to arrange the elements in order of increasing values of atomic numbers rather than their relative atomic masses. This cleared up several problems in the ordering of the elements.

The observations of periodicities of their physical and chemical properties has allowed the elements to be arranged into eighteen groups. This is extremely helpful to chemists, who try to rationalize observations in terms of theories of atomic and molecular structure. Group 1 consists of the elements hydrogen (Z = 1), lithium (Z =

3), sodium (Z = 11), potassium (Z = 19), rubidium (Z = 37), caesium (Z = 55) and francium (Z = 87). Group 18 consists of the elements helium (Z = 2), neon (Z =10), argon (Z = 18), krypton (Z = 36), xenon (Z = 54) and radon (Z = 86). The successive differences between the Z values of the Group 1 elements are identical to those of the Group 18 elements; 2, 8, 8, 18, 18 and 32, these numbers being equal to the number of elements in successive periods. The chemical unreactivity of the Group 18 elements indicates that atoms with 2, 10, 18, 36, 54 or 86 electrons are particularly stable chemically. In elements succeeding any one of the Group 18 members, the excess of electrons over these numbers are regarded as the valency electrons, some or all of which may be used in compound formation. The most important chemical property of an element is its combining power or valency. The valency of an element is dependent upon its number of available valency electrons, this being discussed later in the Chapter. Group 1 elements are monovalent in that they possess one valency electron, they each immediately succeed a Group 18 element of the earlier period, and can combine with one monovalent element in forming a compound. For example, the compound AB is formed from the two monovalent elements represented by the symbols A and B. Group 2 elements possess two valency electrons and are bi-valent in that each atom (A) can form a compound with two monovalent atoms (B) to form the compound AB_2. The common valencies of the elements of the first short period, lithium to neon, (see Fig. 11.3) follow the sequence, one (Li), two (Be), three (B), four (C), three (N), two (O), one (F) and zero (Ne). The sequence of common valency values; 1, 2, 3, 4, 3, 2, 1, and 0, is repeated in the eight elements of the second short period.

1	2	3	4	5	6	7	8	9	10	11	12	13	14	15	16	17	18
1 H																	2 He
3 Li	4 Be		← atomic number									5 B	6 C	7 N	8 O	9 F	10 Ne
11 Na	12 Mg		← element symbol									13 Al	14 **Si**	15 P	16 S	17 Cl	18 Ar
19 K	20 Ca	21 Sc	22 Ti	23 V	24 Cr	25 Mn	26 Fe	27 Co	28 Ni	29 Cu	30 Zn	31 Ga	32 **Ge**	33 **As**	34 Se	35 Br	36 Kr
37 Rb	38 Sr	39 Y	40 Zr	41 Nb	42 Mo	43 Tc	44 Ru	45 Rh	46 Pd	47 Ag	48 Cd	49 In	50 Sn	51 **Sb**	52 **Te**	53 I	54 Xe
55 Cs	56 Ba	71 Lu	72 Hf	73 Ta	74 W	75 Re	76 Os	77 Ir	78 Pt	79 Au	80 Hg	81 Tl	82 Pb	83 Bi	84 Po	85 At	86 Rn
87 Fr	88 Ra	103 Lr	104 Db	105 Jl	106 Rf	107 Bh	108 Hn	109 Mt									
		57 La	58 Ce	59 Pr	60 Nd	61 Pm	62 Sm	63 Eu	64 Gd	65 Tb	66 Dy	67 Ho	68 Er	69 Tm	70 Yb		
		89 Ac	90 Th	91 Pa	92 U	93 Np	94 Pu	95 Am	96 Cm	97 Bk	98 Cf	99 Es	100 Fm	101 Md	102 No		

Fig. 11.3 The modern version of the periodic table of the elements; the metalloid elements are shown in bold. They separate the metals, to their left, from the non-metals to their right. Hydrogen, boron and astatine are non-metals

The modern version of the periodic table is shown in Fig. 11.3. The elements are arranged in horizontal periods in the order in which they appear in Tables 11.2 - 11.6 so that they also form vertical 'Groups', the first period being very short and consisting of the two elements hydrogen and helium. The next two short periods both contain eight elements, lithium to neon, and sodium to argon. Then come the long periods, the first of which contains the eighteen elements from potassium to krypton. Ten elements, from scandium to zinc, called transition elements separate the Groups numbered 2 and 13. The second long period mirrors the first with eighteen elements from rubidium to xenon including the ten transition elements, from yttrium to cadmium, in groups 3 to 12. The third long period is similar to the other two, with the addition of fourteen so-called rare earth elements or lanthanides (lanthanum to ytterbium) coming between barium and lutetium, the latter being the first of the third series of transition elements. Both lanthanum and lutetium have claims to be the first member of the third set of transition elements. Some versions of the periodic table have lanthanum as the first transition element of the period which is then followed by the 14 elements from cerium to lutetium as the lanthanides. Other versions of the table sidestep the issue and place all 15 elements from lanthanum to lutetium as a separate set of lanthanides. All 15 elements belong to Group 3 of the table. The nine other transition elements of the third long period are hafnium to mercury, the period being completed with the six elements from thallium to radon. The third long period contains a total of 32 elements. The fourth long period is unfinished and currently consists of only twenty three elements. It begins with the elements francium and radium in Groups 1 and 2 respectively. These are followed by fourteen elements from actinium to nobelium, known as the actinides, which are all members of Group 3 as is the next element lawrencium. Similar arguments apply to the positions of actinium and lawrencium as are described for those of lanthanum and lutetium above. The elements with atomic numbers greater than that of plutonium ($Z = 94$) are all man-made and are all intensely radioactive with short lives.

Each of the groups of the periodic classification consists of elements that possess broadly similar chemical properties which distinguish them from any other group. The modern classification depends upon a greater understanding of atomic structure than was available to the chemists of Mendeleev's time. The sequence of numbers of elements in each period is now understood from the basis of a very sound theoretical background. Mendeleev's contribution was very important for the further development of our understanding of the elements. He was a famous and well respected man. One of his assistants, called Shostakovich, named his son Dimitri as a mark of respect for Mendeleev. Dimitri Shostakovich found fame as the greatest composer of music in the twentieth century.

The general structure of the periodic table can be described in terms of four blocks which are named by the code letters, s, p, d and f, derived from the theory upon which the structure is based. Fig. 11.4 shows a periodic table divided into the four blocks of elements. Groups 1 and 2 plus helium from Group 18 form the elements of the s block, Groups 3 to 12 contain elements of the d block, i.e. the transition elements. Also classified in Group 3 are the elements of the f block, i.e. the rare earths or lanthanides, and the actinides. The elements of Groups 13 to 17 plus the elements of Group 18, except for helium, form the p block. Some treatments of chemistry of the elements classify the s and p block elements as main group elements, the remainder being the d block transition elements and the f block lanthanides and actinides. Of the 109 elements that either have a natural existence or have been synthesized, 85 are metals, 6 are known as metalloids because they have properties which are intermediate between metals and non-metals, and only 18 are non-metals (see Fig. 11.3). With regard to their physical state at 25°C and at one atmosphere pressure, so-called standard conditions, eleven of the elements are gases, two are liquids (bromine and mercury) and the great majority of ninety six are solids. There is a tendency for a change from metallic to non-metallic character as the table is traversed from left to right across any period. Down each Group there is a change towards more metallic character, Group 13, i.e. non-metallic boron to metallic aluminium, gallium, indium and thallium, and Group 14, i.e. non-metallic carbon, metalloid silicon and germanium, and metallic tin and lead, being good examples of the tendency. Metalloids exhibit semi-conductor properties, silicon, germanium, arsenic and selenium being important in the electronics industries.

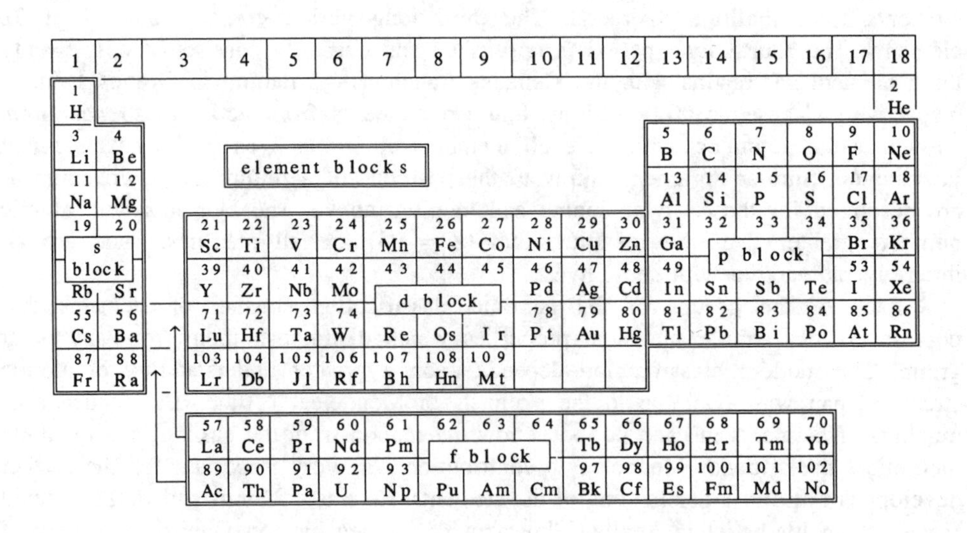

Fig. 11.4 The s, p, d and f blocks of the periodic table

11.5 MOLECULES AND COMPOUNDS

The molecular structures of some elements. Electronegative and electropositive elements. Covalent and metallic bonding. Small molecules and giant arrays. Ionic compounds.

The main principles of molecule formation may be exemplified by reference to a selection of elements and compounds. Molecules of elements contain identical atoms, whilst those of compounds contain at least two different elements. The formation of chemical bonds between atoms of elements depends upon their sharing one or more of the available valency electrons.

The formation of the gaseous diatomic molecule of hydrogen, H_2, may be understood in terms of the sharing of electrons between the two atoms. An atom of hydrogen possesses a single electron. If two hydrogen atoms share their two electrons an electron-pair bond is formed. Such bonding is described as covalent. The two negative electrons are situated mainly between the two positive protons and confer stability upon the molecule. The single covalent bond of the hydrogen molecule is the strongest of all such bonds found between atoms of the same element. The molecule has two nuclei, but is otherwise similar to the atom of helium in that both H_2 and He contain a stable pair of electrons. The helium atom has no tendency to undertake any chemical interaction with any other atom because of the presence of the two electrons and exists as a monatomic gas.

The elements of the first short period use their valency electrons, which vary from one to eight in a regular manner, in bond formation. The standard states of the elements vary along the period, lithium and beryllium being solid metals, boron and carbon being solid non-metals, nitrogen, oxygen and fluorine being gaseous diatomic molecules, and neon being a monatomic gas.

The atom of lithium, the first metallic member of Group 1, has a single valency electron as does atomic hydrogen, but prefers to form an infinite three-dimensional array of lithium atoms which allows the element to be a metal. The atoms are arranged in the body centred cubic (bcc) manner as is shown in Fig. 2.2 and by the photograph of the iron structure in Fig. 2.3.

An atom in a metallic structure has too many neighbours to allow covalent bond formation to occur. It is not possible for any atom in such an environment to form conventional electron pair covalent bonds with all its neighbouring atoms, the number of electrons being insufficient. Metallic bonding consists of all the atoms in a piece of metal sharing all their valency electrons. Instead of the electrons being localized in particular molecules they are *delocalized* throughout the whole metallic structure. This is consistent with the well known high electrical conductivity shown by metals, that phenomenon being a good indicator of the metallic state. In the conduction of electricity, i.e. the passage of a stream of electrons along a metal wire, an electron may join the wire at one end at the same time as another electron leaves from the other end. Materials in which the chemical bonding is localized within each molecule are insulators, having very poor electrical conductivities.

Carbon, the first element of Group 14, is the most important of all elements as it is the basis for all organic compounds including, of course, the substances of which human beings are mainly composed. The outstanding feature of carbon compounds is the presence of chains of carbon atoms of various lengths. The element exists naturally as one or other of two well known crystalline forms, diamond and graphite. Diamond is, in its purest form, a very hard colourless crystalline material which is an electrical insulator and definitely non-metallic in character. Graphite is a soft black crystalline material, which is used as a lubricant and in the manufacture of pencils, and is a reasonably good conductor of electricity, although its conductivity, which is about 0.1% of that of copper metal, is around the bottom of the metallic class. Its conductivity, however, is far higher than the silicon, germanium and gallium arsenide semi-conductors that are the basis of the present day electronics industries. The conductivity of graphite is at least a factor of ten thousand better than that of silicon.

The carbon atom has four valency electrons which are all used in a tetrahedral disposition in the structure of diamond as shown in Fig. 11.5 in which the full lines joining the atom positions may be regarded as covalent bonds. The dotted lines are included in the diagram to indicate that the structure of diamond is based upon a cubic system and have no relevance to bonding. The use of lines to join up atoms in structural diagrams can be confusing. Normally they are used to help the reader to visualize the three-dimensional arrangements of atoms and have no other significance.

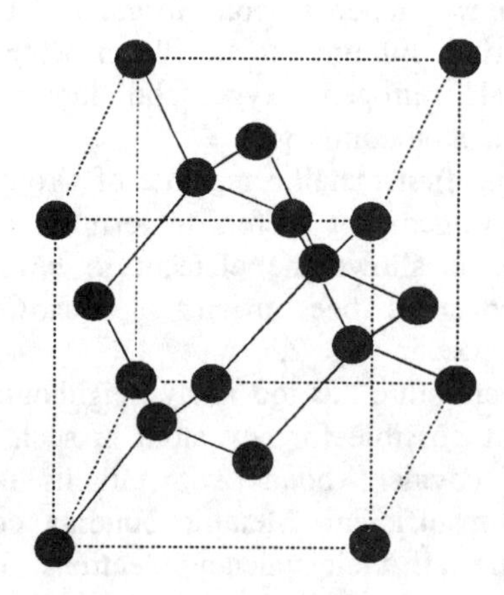

Fig. 11.5 The structure of the diamond form of carbon; the diagram shows the essential cubic nature of the diamond structure and the tetrahedral environment of the carbon atoms. The atoms in the face-centres and at the corners are shared by adjacent cubes in the 'infinite' arrangement of a crystal

In the diamond structure, shown in Fig. 11.5, there is one carbon atom in the centre of each of the six cube faces. The four carbon atoms within the cube are surrounded tetrahedrally by four other carbon atoms, one of these being at a cube corner and the other three being in face centres. The tetrahedral nature of the diamond structure is demonstrated by the photograph in Fig. 11.6 shows the De Beers diamond research laboratory in Johannesburg. The diamond structure is an example of an infinite array of covalently bonded atoms. The valency electrons are all used in the formation of a very stable material. The single carbon-carbon bond is fairly strong (the C-C bond energy is 346 kJ per mole), the participation of each carbon atom in four such bonds giving diamond its exceptional stability. Diamond is chemically very inert. If diamond is heated in the absence of air, it is transformed into graphite, a very expensive experiment to perform, which indicates that the latter is the more stable form.

Graphite consists of two-dimensional infinite sheets of carbon atoms arranged hexagonally so that each carbon atom is covalently bonded to three neighbouring atoms. The general structure of the sheet is reminiscent of that of chicken wire and is shown in Fig. 11.7, together with an indication of the relative position of the adjacent sheets as shown by the dotted lines.

Fig. 11.6 A photograph of the De Beers diamond research laboratory in Johannesburg showing the tetrahedral environment of carbon atoms in the diamond crystal [Photo: E.Barrett]

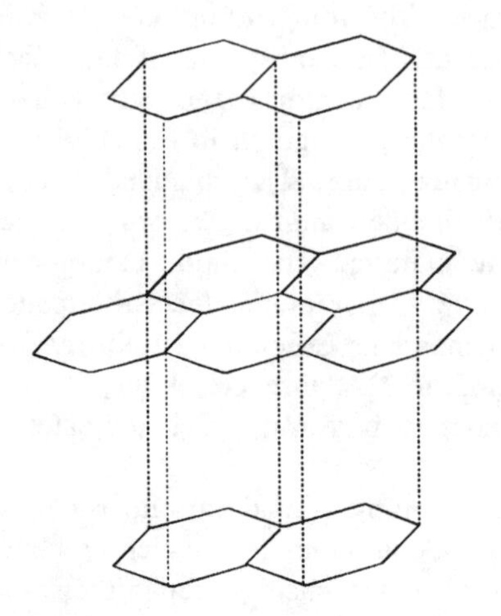

Fig. 11.7 The structure of the graphite form of carbon; the diagram shows small sections of three 'infinite' layers or sheets of hexagonally arranged carbon atoms and their relative positions. Graphite crystals consist of an 'infinite' number of sheets assembled according to those indicated in the diagram

As is usual with carbon structures, the atoms are not indicated; just their positions by the junctions of three lines (covalent bonds). The sheets are piled up to make the three-dimensional structure but the distance between the sheets (335 picometres) is considerably longer than the distance between neighbouring pairs of atoms within one sheet (142 picometres). In general, the longer a bond is, the weaker it is. The inter-sheet bonding is weak and so allows the sheets to slide over each other easily and accounts for the lubricant properties and softness of graphite. The weak bonding allows pencils to work. One valency electron per atom is delocalized in the structure of the graphite sheets and this confers the electrical conductivity upon the material. A single crystal of graphite exhibits conductivity along directions parallel to the sheets but is an insulator with regard to the direction at right angles to the sheets.

Charcoal, produced by burning wood with a restricted amount of air, is a micro-crystalline, or amorphous, form of carbon in which the atoms are arranged locally as they are in graphite but there is no long-range order. It is useful for drawing and as an adsorbant, many molecular species being capable of being adsorbed between the planar parallel layers.

A carbon molecule of current interest is a cluster containing 60 atoms, C_{60}, the structure of which is exemplified by some modern soccer balls; those which are constructed from pentagonal and hexagonal panels. The connectivities of the carbon atoms are similar to those in the sheets of graphite but with each hexagonal arrangement participating in a five-membered carbon ring. The molecule is sometimes referred to by the name, buckminsterfullerene, after the American architect,

Buckminster Fuller, who designed geodesic domes with similar constructions to that of the C_{60} molecule. Its proper chemical name is hexacontacarbon.

The next three elements of the first short period are non-metals and all form gaseous diatomic molecules. The first member of Group 15 is nitrogen with its five valency electrons. Because two of the electrons are normally paired up and cannot be unpaired without causing another pair to be formed, the valency of nitrogen is normally three. The element exists in the gaseous state and forms 78.09% by volume of dry air in the form of diatomic molecules, N_2. Because the molecule is diatomic and its expected valency is three it is considered that the two atoms are bonded together by a triple bond, one that consists of three pairs of shared electrons. The bond is very strong, the energy required to break it being over twice that needed to break the single bond in the hydrogen molecule. The high bond strength causes the nitrogen molecule to be fairly unreactive. Thunderstorms cause some reaction of dinitrogen with dioxygen to give oxides of nitrogen including nitric oxide, NO, and nitrogen dioxide, NO_2, the energy of the lightning discharge providing the necessary energy to break the strong N≡N triple bond. Oxygen atoms generated in the atmosphere by the effect of ultra-violet radiation or thunderstorms on dioxygen can react with dinitrogen to give dinitrogen oxide, N_2O, without the breakage of the nitrogen-nitrogen bond. A very slight amount of nitrogen oxide formation takes place in the hot exhausts of vehicles propelled by internal combustion engines. Together with other material ejected by exhaust pipes, this can offer a slight environmental threat. Pollution of the atmosphere by nitrogen oxides is discussed in Chapter 4. Although the trivalent nitrogen atom has the theoretical possibility of participating in a three-dimensional array, the great strength of the N≡N triple bond, compared to the relatively weak N-N single bond, precludes such structures.

Oxygen, the first member of Group 16, is an element vital to our existence, forming 20.95% by volume of dry air as diatomic dioxygen molecules, O_2. The atom is normally divalent, the strength of the bonding between the two oxygen atoms being quite high. Dioxygen is fairly reactive, especially at elevated temperatures, as is known from the difficulties of controlling fires. Human and animal life in general is dependent upon the oxidizing power of dioxygen to maintain energy supplies by the metabolism of food. The triatomic molecule ozone, O_3, properly called trioxygen, is also crucial to our existence as it shields us from the stream of more energetic photons in which the Earth is bathed by solar radiation. This point is discussed fully in Chapter 4.

Fluorine is the first element in Group 17 and in elemental form exists as the diatomic difluorine molecule, F_2. The molecule is quite weakly bonded and so is very reactive. Its reactivity provides the reason for its discovery and preparation being as late as 1886. It was prepared by Henri Moissan who was awarded the Nobel Prize for chemistry in 1906. The weak bonding is consistent with the pairing of the seven valency electrons of each atom to give three pairs, leaving only one unpaired electron per atom to participate in bond formation.

The last group in the periodic table is Group 18, the member of the first short period being neon. The eight valency electrons are completely paired up, so no sharing with any other atom is possible. The element is monatomic and exists as a gas, to the extent of 0.0018% by volume of dry air, under standard conditions.

The above survey of some of the eight elements of the first short period gives some insight into the diverse nature of elemental properties, the other periods showing some similarities, group for group, but also showing quite large differences due to the trend down any group, which is towards more metallic behaviour. Some examples of these variations are given by boron and aluminium in Group 13, nitrogen and phosphorus in Group 15, and oxygen and sulfur in Group 16. Boron is a non-metal, aluminium is metallic. Elemental nitrogen exists as gaseous dinitrogen, N_2, but phosphorus is a solid consisting of tetrahedral P_4 molecules, in its yellow form. Elemental oxygen exists as gaseous dioxygen, O_2, but sulfur is a solid consisting of S_8 molecules.

The nature of chemical compounds is dependent upon the regions, or blocks of elements, of the periodic table that contain the combining elements. The metallic elements at the left hand side of the table have a tendency to form positive ions by losing one or more electrons. Such elements are described as electropositive. The non-metallic elements at the right hand side of the table have a tendency to form negative ions by gaining extra electrons. They are the electronegative elements. There is a general trend across (left to right) any periodic group from electropositive to electronegative character with some of the intermediate elements showing little tendency to form ions. Compounds formed between two elements that differ very little in their ion-forming properties are either metallic alloys (formed by two metallic, electropositive elements from the s, d or f blocks) or are covalently bonded. When an electropositive metal reacts with an electronegative non-metal one or more electrons are transferred from the electropositive element to the electronegative one, an ionic compound being formed. The most familiar example of an ionic compound is that formed from sodium and chlorine, sodium chloride (common salt) with the formula NaCl. The chemical equation which represents the reaction is written as:

$$2Na + Cl_2 \longrightarrow 2NaCl$$

Although the formula of sodium chloride is very simple, it does not imply that the compound consists of diatomic molecules. The solid substance consists of positively charged sodium ions, Na^+, and negatively charged chloride ions, Cl^-, arranged in a cubic fashion as shown by the diagram in Fig. 11.8.

The ions in the diagram are shaded differently to distinguish the two different ions; Na^+ and Cl^-. Which is which does not matter since both ions have an identical arrangement in the infinite structure. Crystals of table salt exhibit the essential cubic nature of the structure. In solid sodium chloride, each ion is surrounded by six ions of opposite charge and the solid is stabilized by the resulting electrostatic attractions which balance the repulsions between like charged ions. When solid sodium chloride dissolves in water to form an aqueous solution, the structure breaks up to give separate sodium and chloride ions which are all separately surrounded by small numbers of water molecules. These aquated or hydrated ions are responsible for the electrical conductivity of an aqueous solution of sodium chloride, pure liquid water being a poor conductor.

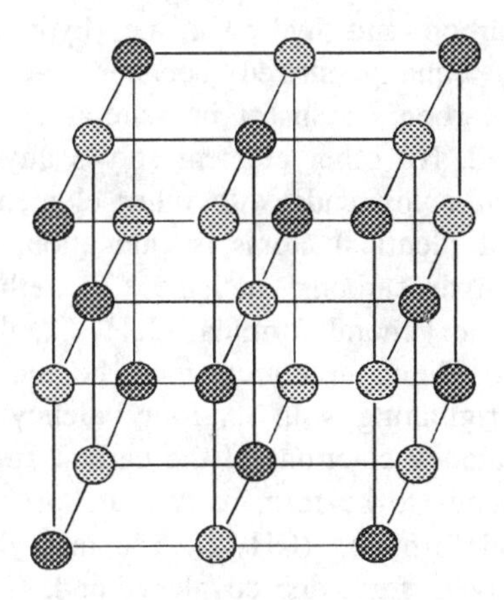

Fig. 11.8 The structure of sodium chloride; the different shading distinguishes sodium ions from chloride ions (it is irrelevant which is which); ions at the corners, along the cube sides and in the face-centres are shared by adjacent cubes in the 'infinite' arrangement of a crystal

Although it is quite correct to ascribe a valency of one to both sodium and chlorine when they combine to form sodium chloride, it is obvious from the crystal structure of the salt that each sodium ion is equidistant from six chloride ions and each chloride ion has six sodium ions as its nearest neighbours. It is not correct to conclude that the atoms are participating in six bonds and that they are therefore hexavalent. Ionic bonding is a long-range phenomenon and a better concept to use in the description of ionic compounds is the oxidation state of each participant. The oxidation state of any neutral atom is taken to be zero. When an ion is formed by the loss of electrons, which is known as oxidation, the ion has an oxidation state equal to the number of electrons lost by the neutral atom. In the case of sodium forming sodium chloride, the neutral atom loses one electron in becoming the uni-positive sodium ion. The oxidation state of the sodium is then said to be +1. Likewise the oxidation state of chlorine in sodium chloride is -1, because the chlorine atom has acquired a negative electron. In both cases the oxidation states are equal to the charges upon the ions, but this is not always so. When uranium metal dissolves in nitric acid the ion UO_2^{2+} is formed. The uranium is bonded to the two normally di-valent oxygen atoms and could be said to be exhibiting a valency of four, but this is not a very helpful description. If the oxidation state of the oxygen is taken to be its normal value of -2 (as in the oxide ion, O^{2-}), that of the uranium must be +6 to allow for the overall charge of +2 on the ion. It is usual to describe the oxidation states of elements by Roman numerals as superscripts to the element symbol, the UO_2^{2+} ion being an example of U^{VI} (i.e. the ion may be regarded as a mixture of U^{6+} and $2O^{2-}$), Na^+ and Cl^- being written as Na^I and Cl^{-I}.

Compounds of carbon are generally covalent molecules. For example, the compounds containing carbon and hydrogen, i.e. hydrocarbons, are covalent, the bonding electrons being equally shared between the participating atoms. The distinguishing feature of carbon chemistry in general is the ubiquity of the carbon - carbon (C-C) single bond. No other element shows anything like its capacity and extent to bond to itself in compounds with other elements. The technical term for the formation of chains of identical atoms is catenation, meaning chain production. In the series of saturated hydrocarbons, methane, CH_4, ethane, C_2H_6, propane, C_3H_8, and butane, C_4H_{10}, with the general formula, C_nH_{2n+2}, the maximum value of n is observed to be seventy. Their structures may be understood in terms of the tetravalence of carbon, originating with its four valency electrons. In a methylene group, CH_2, two of the valency electrons of the carbon atom are used, with the other two allowing two more bonds to be formed. A chain of n methylene groups may be formed, giving the general formula, $-(CH_2)_n-$. The methylene groups at the ends of the chain have the capacity to form one covalent bond. If these two bonds are made with hydrogen atoms stable molecules are formed with the general formula, $H-(CH_2)_n-H$, i.e. C_nH_{2n+2}.

```
 H H          H H H
 | |          | | |
H-C-C-H     H-C-C-C-H
 | |          | | |
 H H          H H H
```

$$H-(CH_2)_n-H = C_nH_{2n+2}$$

Fig. 11.9 Structural formulae of ethane, propane and the generalized formula of saturated hydrocarbons

This way of writing the formula implies that the hydrocarbon series consists of so-called straight chain compounds as shown in Fig. 11.9, although all the bond angles are about 109°.

A phenomenon known as isomerism (introduced in Chapter 2) is widespread in carbon - hydrogen chemistry and is exemplified by the two structural isomers of butane shown in Fig. 11.10 which shows the bonding in the straight chain molecule and in the branched isomer which is called iso-butane or by its proper name 2-methylpropane, a propane molecule with one of the hydrogen atoms on the second carbon atom (hence the 2 in the formula) being replaced by a methyl group.

$CH_3CH_2CH_2CH_3$

```
CH3CHCH3
    |
   CH3
```

Fig. 11.10 The structural formulae of butane and its isomer, iso-butane or 2-methylpropane (the proper name, derived from propane by replacing a hydrogen atom on its second carbon atom with a methyl group, CH_3)

The number of possible isomers increases as the value of n increases. There are eighteen possible octanes and over 62 million million possible isomers of the C_{40} hydrocarbon!

Because carbon is tetravalent, many of its compounds contain carbon-carbon double or triple bonds in which the adjacent pair of carbon atoms share four or six electrons respectively. The simplest examples of these so-called unsaturated compounds are ethene (commonly known as ethylene), C_2H_4, and ethyne (commonly known as acetylene), C_2H_2. The formulae of these compounds may be written as $H_2C{=}CH_2$ and $HC{\equiv}CH$ to indicate the multiplicity of the carbon-carbon linkages. There are many hydrocarbons, known as aromatic compounds, which are based upon the molecule of benzene, C_6H_6. Benzene is a planar, regularly hexagonal, molecule as shown in Fig. 11.11, the bond angles all being 120°.

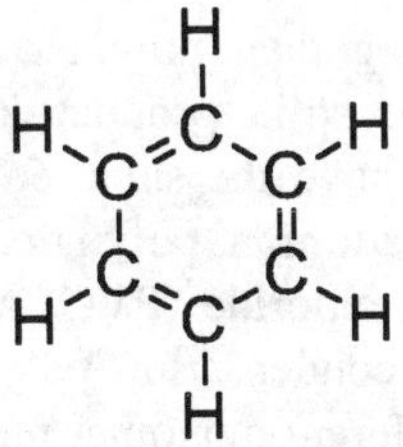

Fig. 11.11 The structure of the benzene molecule showing the positions of the hydrogen atoms

The bonding of the carbon hexagon is a mixture of localized C-C single bonds, superimposed on which is the extra bonding from six delocalized valency electrons. The same delocalized electrons are used in the inter-sheet bonding in graphite, the C-H bonds of benzene being equivalent to the intra-sheet C-C bonds of graphite. The extra bonding stabilizes the molecule, causing it to have different properties from those compounds with localized C=C double bonds. It is normal to omit the hydrogen atoms in diagrams of the structural formulae of carbon compounds. The two usual methods of writing the structural formula of benzene are shown in Fig. 11.12, including the regular hexagon with an inscribed ring to indicate the delocalized electrons.

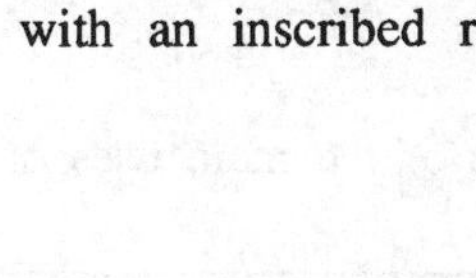

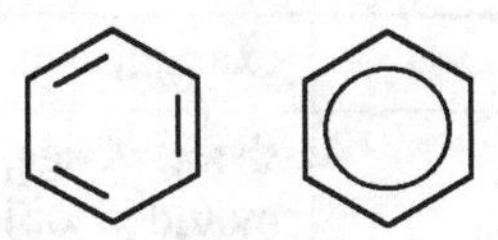

Fig. 11.12 Alternative methods of writing the structure of benzene showing the carbon hexagon but omitting the hydrogen atoms; the formula on the left indicates alternating single and double carbon-carbon bonds, the one on the right implying that the single and double bond characters are delocalized over the regularly hexagonal molecule

One very important industrial aspect of carbon chemistry is the production of polymers. There are many ways in which some small carbon compounds may be polymerized. The small contributing molecules are called monomers and can be polymerized in two general ways, known as addition polymerization and condensation polymerization. The simplest example of addition polymerization is the production of polythene from ethene (ethylene). The combination of ethene molecules to form a polythene chain may be written as the general equation:

$$n(CH_2{=}CH_2) \longrightarrow (CH_2\text{-}CH_2)_n$$

many molecules of ethene joining together (polymerizing) to give long polymer chains which, jumbled together, form the solid product known as polythene. In addition polymerization, the product chains have the same composition as the monomers from which they are composed. Other addition polymers are polypropylene, polystyrene, polyvinyl acetate (PVA), polyvinyl chloride (PVC) and polytetrafluoroethene (PTFE).

There are two main forms of condensation polymers, both being produced when water is eliminated between two different monomer molecules. Polyamides are produced by the condensation of an acid, e.g. terephthalic acid, $HOOCC_6H_4COOH$, with two acidic COOH groups per molecule, with a diamine, e.g. 1,6-diaminohexane, $H_2N(CH_2)_6NH_2$ (with amino-groups, NH_2, attached to carbon atoms 1 and 6 at the ends of the six-carbon chain). The elimination of water between the two monomers arranged in an alternating sequence produces the condensation polymer known as the polyamide, 6,6-nylon, because both monomers possess six carbon atoms. The condensation reaction is identical to that which occurs when proteins are synthesized from amino-acids and is described in Chapter 9. If a diol, i.e a double alcohol containing two OH groups, is used in place of the diamine in the condensation reaction with the di-acid the product is a polyester, Terylene™ and Crimplene™ being common examples. Other examples of carbon compounds feature in Chapters 6, 9 and 10. Some of the main applications of addition polymers are given in Table 11.7.

Table 11.7 - Some of the main uses of addition polymers

polymer	examples of main uses
polythene	sheets, bags, tubing, bottles
polypropylene	moulded objects, carpet fibres
polystyrene	rigid foams, ceiling tiles
polyvinylacetate	paints, adhesives
polyvinylchloride (PVC)	gramophone records, clothing
polytetrafluoroethene	non-stick pan linings
polymethylmethacrylate	Perspex™
polyacrylonitrile	Orlon™ fibres

Just as there are differences in the properties of the elementary forms of the elements in any Group, there are differences in their compounds, some slight and some very profound. The Group 1 hydrides (symbolized as M^+H^-) are very similar to each other, both in structure and in their reactions. They are all solid ionic compounds and react with water to produce gaseous dihydrogen and alkaline solutions:

$$M^+H^-(s) + H_2O(l) \longrightarrow M^+(aq) + OH^-(aq) + H_2(g)$$

The oxide of silicon, found in nature as quartz, has the formula, SiO_2, and although having a formula of the same form as that of carbon dioxide, could not be more physically different from it. Whereas carbon dioxide is a discrete gaseous triatomic molecule, which is linear with the carbon atom being central, under standard conditions, silicon dioxide is a high melting solid (m.p. 1723°C) with a giant three-dimensional structure. The significant difference between carbon dioxide and silicon dioxide lies in the bond strengths, the silicon-oxygen single bond being particularly strong. The structure of silica may be understood from the basis of a tetrahedral diamond-like arrangement of the silicon atoms with every pair of silicon atoms being bridged by an oxygen atom. The associated silicic acid, H_4SiO_4 or $Si(OH)_4$, is the basis of the understanding of the vast number of silicate minerals which constitute the majority of the Earth's crust. The tetravalency of silicon allows for the production of single chains of $-(O\text{-}SiO_2)_n-$ linkages which are contained in compounds with the formula, M_2SiO_3, M representing any monovalent metal. Single chains may be cross-linked to give compounds with a repeating unit which is a double chain with the formula unit, $-Si_4O_{11}-$. Double chains are found in minerals with the formulae, $M_4Si_4O_{11}$, asbestos being an example. If an infinite number of chains are joined together to give a two-dimensional sheet the formula of the minerals becomes $M_2Si_2O_5$, mica being an example. The fibrous nature of asbestos and the sheet-like appearance of mica are consequences of their structures, consisting of infinite double chains and infinite two-dimensional sheets respectively. If the sheet-like structures become cross-linked the resulting infinite three-dimensional solid is silicon dioxide with the simple formula, SiO_2.

As any group of elements in the periodic table is descended, the elements become more electropositive, less electronegative, the elements becoming less non-metallic and more metallic in character. Group 14 exemplifies that statement. As the group is descended there is a gradual transition from non-metallic carbon and silicon, metalloid germanium to metallic tin and lead. There is a transition from the long-chain molecules formed by carbon atoms, upon some of which life is dependent, to the aqueous lead(II) ion, $Pb^{2+}(aq)$, which is poisonous to humans and other animals.

The main points concerning acids and bases are contained in Chapter 2. Whether a given compound acts as an acid, i.e. ionizing to give hydrogen ions, or a base i.e. ionizing to give hydroxide ions, in aqueous solution, can be understood in general terms by considering the molecule R-O-H, the R group representing the remainder of the molecule apart from the OH group. If the O-H bond is weaker than the R-O bond it is likely that the compound will act as an acid and ionize in aqueous solution to give hydrogen ions and RO^- ions. Alternatively, if the O-H bond is stronger than the R-O bond it is the latter which is likely to break and give R^+ ions and hydroxide ions. The R-O bond strength is usually low if the R group is electropositive and large, e.g. R = Na, but is high if the R group is electronegative and small, e.g. R = HSO_3 as in sulfuric acid, H-O-SO_3H, where the small sulfur atom in oxidation state VI is linked to the OH group.

11.6 WHY CHEMICAL REACTIONS OCCUR

States of matter. Temperature. Entropy and disorder. Criterion of reaction feasibility.

Matter normally exists in one or other of the physical states, gaseous, liquid and solid. A fourth state is a particular kind of gaseous state in which the materials are so hot that molecules dissociate into atoms, and atoms ionize. Atoms, ions and electrons co-exist in this fourth state which is called plasma and is relevant to nuclear processes that occur in the Sun and to very high temperature systems that can be produced in the laboratory and in thermonuclear explosions. The states of matter vary considerably in the order or disorder exhibited by the arrangements of the atoms, ions or molecules of which they consist.

There are four kinds of solids with highly ordered structures; metals, non-metallic covalently bonded giant structures, ionic crystals, molecular crystals, and other solids, such as plastics and glass, which have disordered structures. The units of the metallic state are atoms, which are arranged in very orderly and simple ways such as the displays of oranges at the fruit shop. The atoms are held together by metallic bonding which has a range of strengths depending mainly upon the number of valency electrons possessed by each atom. The Group 1 metals have a single valency electron and have relatively weak bonding as is clear from their low melting points, e.g. the melting point of cesium is only 302 K. The metal with the highest melting point is tungsten (m.p. 3680 K) which has six valency electrons available for bonding.

Examples of structures which consist of giant arrays of covalently bonded atoms are diamond, silicon carbide (SiC, carborundum) and aluminium oxide (Al_2O_3, corundum). The materials all have very high melting points, the previous examples having melting points of 3820, 2700 and 2072 K respectively. These high values are indicative of the high strengths of the covalent bonds linking the atoms together in three-dimensional networks.

Ionic crystals are giant arrays of regularly arranged positively and negatively charged ions. The electrostatic attractions balance the repulsions to produce very stable solids with high melting points, e.g. sodium fluoride (1266 K) and calcium fluoride (1696 K). There are many solids with ionic-type crystal structures that strictly should not be regarded as fully ionic. Such compounds possess a considerable fraction of covalent character and have fairly low melting points, e.g. copper bromide, m.p. 771 K. Ionic solids are generally soluble in water, the solutions containing aquated ions which contribute to electrical conduction. Ionic compounds also conduct electricity in their molten state, which is consistent with their being composed of ions.

The fourth type of ordered solid consists of compounds composed of molecules which form molecular crystals. The units of the crystals are the individual molecules and these are held together by relatively weak intermolecular forces. They have low melting points in consequence, and melt to give liquids composed of the less ordered molecules. Molecular crystals vary from the solid formed by helium atoms at temperatures below 3.5 K to aspirin which melts at 408 K. There are many examples of compounds which have melting points as high as 550 K and there is some overlap with the values for compounds that are normally regarded as ionic.

A liquid takes up a position in a container which is consistent with gravity and has a viscosity which allows translational movement of the units of the substance of which it consists. A liquid has a fixed volume at a given temperature and pressure with the molecules moving around at random within it. The molecules move at variable speeds which change as they collide with each other and with the sides of the container. There is considerable disorder in a normal liquid, but if it were possible to take a 'snapshot' it would show a semblance of local order, a different arrangement being observed with every snapshot. It is as though the material is partially crystalline in very small regions, there being a constant interchange between the regions. There is some space available in the liquid state to accommodate solute molecules. When a solute is dissolved in a solvent to make a solution there is often a change in volume of the liquid phase. If the solute molecules are large they force the solution to have a greater volume than that of the solvent, the molecules of which are then further apart. If an ionic compound dissolves in water it is possible for there to be a contraction because the ions attract a small number of water molecules to themselves and force the water molecules to be more tightly bound in one region than they are in the pure liquid.

A gas takes up all the space it can find and has to be contained in an air-tight vessel for proper study. The molecules, or atoms in the case of the group 18 elements, move around at random but, like those in the liquid phase, they have a wide distribution of speeds. The speed of any one molecule changes as collisions with other molecules and the sides of the vessel occur.

The understanding of both the gaseous phase and temperature are closely linked. The temperature of a gas is proportional to the average speed of its molecules. The faster the molecules move, the higher is the perceived temperature, either by human skin or a suitable thermometer. The same is true of liquids, the molecules of which have similar translational motion, i.e. movement in three dimensions to any part of the occupied volume, as do gaseous molecules. The constituent units of solids have no translational freedom but, in aggregate, they have a very large number of possible vibrations and it is the transfer of the vibrational energy to nerve endings which allows our perception of the temperature. Independently of composition, hot bodies, left to themselves, cool down. The reverse process never happens spontaneously. Newton noticed that hot objects cooled down and enunciated his law of cooling. This is that the rate of cooling of an object at a temperature, T, is proportional to the difference in temperature between T and room temperature which implies that the hotter a body is, the greater is its rate of cooling, but the rate slows down as its temperature approaches that of its surroundings.

The transfer of heat energy between bodies is important in the understanding of heat engines. Its study has formed the basis of the subject known as thermodynamics. The subject has very important applications in the understanding of why chemical reactions occur. There are three laws of thermodynamics. The first law is sometimes known as the law of conservation of energy, which is that 'energy cannot be created or destroyed, but can change from one form to another'. Chemical energy can be transformed into heat, e.g. by burning coal or other fuels, or into electrical energy as occurs in a battery. The second law informs us that spontaneous changes, i.e. the ones that take place without any interference, are associated with an increase in disorder or untidiness. This law is discussed in detail later in the chapter. The third law is almost self-evident and tells us that a perfect crystalline substance has zero disorder. The three thermodynamic laws may be paraphrased in the following manner.

Law I. Nothing is free.

Law II. Drop things and they will probably break. The pieces never re-assemble without work being done. It is the basis of Murphy's law.

Law III. Nothing is perfect.

Most chemical substances found in the home are stable with respect to their component elements. Sodium chloride (NaCl) is more stable than the sodium metal and dichlorine gas from which it may be made and the methane molecule (CH_4) of natural gas is more stable than its component carbon and dihydrogen molecules. In these examples amounts of heat are given out when the compounds are formed, the reactions being described as being exothermic.

Some reactions are accompanied by an absorption of heat and are known as endothermic processes. The formation of ethyne, the formal name for acetylene (C_2H_2), from its constituent elements is endothermic. This implies that the ethyne molecule is less stable than its component elements. This instability of ethyne contributes to its explosive properties when mixed with dioxygen, the oxy-acetylene welder being well aware of the dangers of using such a mixture. When used to provide the fuel for a flame, ethyne produces a very high flame temperature because of the inherent instability of the compound. Only the saturated hydrocarbons are produced from their elements with the emission of heat, this being the reason for the naturally occurring hydrocarbons being very largely of the saturated kind. Saturation, in this context means that the carbon atoms are forming four single covalent bonds to hydrogen or other carbon atoms and are not engaged in the formation of C=C double or C≡C triple bonds as in the molecules of ethene and ethyne respectively. Any double or triple bonding is known as unsaturation and leads to instability in such molecules.

Heat changes are associated with most physical and chemical changes. A possible criterion for the feasibility of a reaction could be the heat change which occurs.

Two experiments demonstrate easily that such a criterion is inadequate. One experiment is to mix roughly equal volumes of water and methylated spirits, which is mainly ethanol with some added methanol to make it undrinkable by sensible people. There is a noticeable increase in temperature which indicates that the process of mixing is exothermic. A second experiment is to dissolve a tablespoonful of salt in a cupful of water. Again the reaction occurs but there is a reduction in the temperature, the reaction being endothermic. In order to restore the temperature to its room value heat has to enter the reacting system, i.e. the solution in the cup. A better criterion is needed which is best introduced by considering the breaking of an egg. The nursery rhyme about Humpty Dumpty and his big fall expresses the matter in a nutshell?! Humpty fell off the wall, moving to a position of lower energy and was transformed from an orderly egg to a scrambled mess on the ground. Nobody, not even all the King's horses and men, could reverse the process. Other everyday experiences are relevant to seeking a good criterion for predicting the feasibility of physical or chemical changes. Houses and office desks seem to become untidy quite spontaneously, they never spontaneously become tidy - that requires work! Crockery usually breaks when dropped or hit, the pieces never joining up again spontaneously. Chemical reactions which occur spontaneously are one-way processes that require much work to be expended to reverse them. There are two tendencies which are helpful in the establishment of the criteria which govern reaction feasibility. One is that reacting systems tend to form products which have lower energy than that of the reactants, endothermic reactions being an exception. The second tendency is for the system to become more disordered. These energy minimization and disorder maximization criteria are the basis for the determination of reaction feasibility.

The technical term for disorder is entropy. The second law of thermodynamics may be stated as 'spontaneous changes are accompanied by an increase in total entropy'. The total entropy of this law includes that of the reacting system and its surroundings, the latter being the rest of the Universe or, in practical terms, the laboratory or room in which the change occurs. This allows for the entropy of the system to decrease, providing that the entropy of the surroundings increases by an even greater amount. The application of the second law may be further discussed for four different cases which depend upon the changes in energy and entropy, both of which can either increase or decrease.

(i) The best conditions for a reaction to be spontaneous are when the reacting system gives out energy to its surroundings, i.e. is exothermic, and becomes more disordered; an increase in entropy. The heat given out to the 'rest of the Universe' causes the entropy of the surroundings to increase, there being no doubt in this case that an overall increase in entropy has taken place. Spontaneous reactions of this kind usually go practically to completion. The reaction of solid, i.e. low entropy, sodium hydroxide with an aqueous solution of hydrochloric acid to give a solution containing sodium and chloride ions plus some extra liquid water is very exothermic and has an associated entropy increase:

$$NaOH(solid) + H^+(aq) + Cl^-(aq) \rightarrow Na^+(aq) + Cl^-(aq) + H_2O(liq)$$

The reaction is one between the hydrogen ion in the solution of hydrochloric acid, $H^+(aq)$ and $Cl^-(aq)$, and the hydroxide ion of the solid sodium hydroxide to produce a water molecule. The sodium ion then has the freedom of movement throughout the resulting solution, i.e. high entropy, whereas in the solid NaOH it had no freedom at all, i.e. low entropy.

(ii) A theoretical reaction which is endothermic, i.e. heat is absorbed from the surroundings which allows the entropy of the surroundings to decrease, and in which the entropy of the system decreases cannot possibly occur spontaneously because there is an overall entropy decrease; the system and its surroundings becoming more ordered in such a case. Such reactions may be made to occur by the expenditure of energy. One example of this case is the endothermic reaction of solid calcium carbonate with an aqueous solution of sodium chloride to give aqueous calcium chloride and solid sodium carbonate:

$$CaCO_3(s) + 2NaCl(aq) \longrightarrow Na_2CO_3(s) + CaCl_2(aq)$$

in which there is a decrease of entropy. There are four aqueous ions on the left hand side of the equation (two Na^+ and two Cl^-) but only three on the right (one Ca^{2+} and two Cl^-). The freedom of movement of one mole of ions is lost in the process, i.e. two aqueous sodium ions are replaced with only one aqueous calcium ion. This non-spontaneous reaction is vital to the alkali industry, i.e. the production of sodium carbonate or soda-ash, and is discussed in detail in Chapter 7. Non-spontaneous reactions can be made to occur only by the expenditure of energy.

In the above two cases the two deciding factors, energy and entropy changes, work in the same direction, but there are two conditional cases in which they are opposed to each other.

(iii) A reaction in which heat is released, i.e. is exothermic, and in which there is a decrease of entropy of the system may be spontaneous provided that the amount of heat released is large enough to cause the entropy of the surroundings to increase by an amount greater than the entropy decrease suffered by the system. This ensures that there is an overall increase in entropy so that the reaction can take place. The general formation of crystals from a solution of a substance exemplifies this case. A solution of a substance has a high entropy, the constituents having the freedom of movement throughout the system. If a solid, i.e. low entropy, separates out the process is usually accompanied by the release of heat into the surroundings, i.e. the process is exothermic. The decrease in entropy of the system is outbalanced by the increase of the entropy of the surroundings caused by the released heat. The process is thus spontaneous.

(iv) Endothermic reactions, in which heat is absorbed from the surroundings, i.e. the entropy of the surroundings decreases, may only be spontaneous if they are associated with a large enough entropy increase to overcome the decrease suffered by the surroundings. The simplest example of this case is the dissolution of solid, i.e. low entropy, ammonium chloride in water to produce an aqueous solution containing aqueous ammonium and chloride ions which have freedom of movement throughout the solution:

$$NH_4Cl(s) \longrightarrow NH_4^+(aq) + Cl^-(aq)$$

The absorption of heat from the surroundings, i.e. the reaction is endothermic, allows an entropy decrease which is overcome by the increase in entropy brought about by the solid compound dissociating into its constituent ions which become aquated in the aqueous solution.

In the two conditional cases, (iii) and (iv), the reactions, although spontaneous, may not all go to completion because of the opposing effects of energy and entropy. In finely balanced cases an equilibrium state is eventually attained in which some of the reactant molecules co-exist with those of the products.

11.7 HOW CHEMICAL REACTIONS OCCUR

Molecular collisions. Range of reaction rates. Effect of temperature on rate. Activation energy. Nitrogen fixation. Production of ammonia by the Haber-Bosch process and by microorganisms. Catalysis by metals and enzymes. Catalytic converters. Cooking. Oxidation of the human body.

The majority of chemical reactions occur in systems which contain at least one component which is either a gas or a liquid, the latter possibly being a solution of a substance in a solvent. It is rare for reactions to take place between solids unless high temperatures are used. The reason for this is that reactions occur when suitable collisions between the reactants take place. In the case of solid explosives a detonator is used to supply the initiating high temperature required to cause the reaction to proceed. Molecules in the gas, liquid and solution states have translational freedom to collide with each other and if the colliding molecules have sufficient energy a chemical reaction occurs. In a one-stage process, such as the reaction between compounds A and B (the reactants) to give compounds C and D (the products):

$$A + B \longrightarrow C + D$$

collisions between A and B are necessary for the reaction to occur. Collisions between A and A or between B and B do take place, any exchange of energy affecting the individual speeds of the participants but not leading to reaction. Not all A-B collisions necessarily lead to reaction because there is usually an energy barrier that the colliding molecules have to surmount before reaction can occur. The energy change as reactants change into products is shown in the diagram of Fig. 11.13.
The diagram demonstrates the difference between thermodynamics and kinetics, thermodynamics being concerned with the initial and final states of a system, i.e. the reactants and products respectively, the passage from reactants to products

being the province of kinetics. The energy barrier is known as the activation energy.

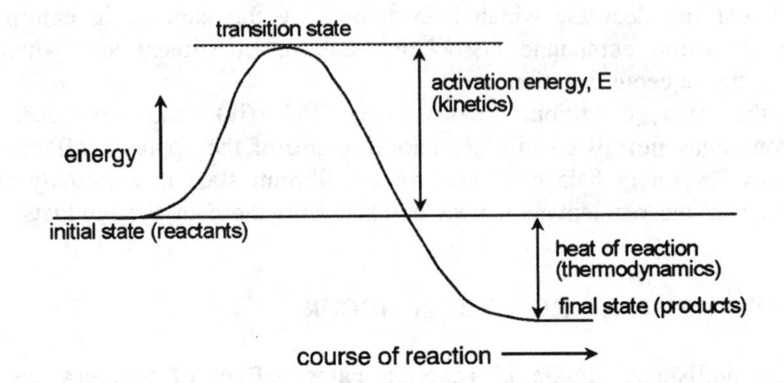

Fig. 11.13 A diagram showing the change in energy of a reacting system as the reaction proceeds; it also shows the energy changes which govern the rate (i.e. kinetics) of a reaction (the activation energy) and the thermodynamics (i.e. the heat change) of the reaction; the two are not connected

It is economically and industrially important to be able to control the rates of chemical reactions. One method of increasing the rate of a reaction is to increase the temperature of the system. A rough rule for the effect of temperature on reaction rate is that the rate doubles for a 10°C rise, although there are many exceptions. The influence of an increased temperature on a reacting system causes the molecules to have a different distribution of speeds, and consequently different kinetic energies. The effect of a ten degree rise is to allow roughly twice the number of collisions between the reactants to have sufficient activation energy to allow the crossing of the energy barrier between reactants and products. In cookery, the temperatures used depend upon the method, food being either boiled in water at atmospheric pressure (100°C) or in a pressure cooker (up to 120°C), baked, roasted or grilled (150-250°C) or heated in oil (up to 300°C). All these methods increase the rate at which the treated foodstuff becomes suitably edible.

The use of higher temperatures to increase the rates of reactions requires the use of more power to provide the extra heating. Apart from the economic cost of this, there are many reactions which cannot be carried out at elevated temperatures, e.g. enzyme reactions in which the enzyme molecule becomes denatured at high temperature. The chemical disadvantages of using higher temperatures are the decomposition of reactants or products and the predominance of unwanted by-products. A much more advantageous method of increasing the rate of a reaction is to use a catalyst.

A catalyst is defined as a substance which alters the rate of a chemical reaction without being consumed in the process. The catalyst participates in the reaction mechanism but, is either unchanged or is regenerated by the process. The more common catalysts are solid surfaces and enzyme molecules. Both kinds of catalyst operate in much the same manner by allowing the adsorption of the reactant molecules to occur. While kept in close proximity the molecules have a much greater chance of reacting with each other than if both reactants had the complete independent translational freedom in the gas or liquid states. The catalyst offers a reaction path which has a lower activation energy than the uncatalysed path. This is also because of the interaction between the catalyst and one or other (or both) of the reactants. One reaction which is used extensively to produce ammonia for the fertilizer industry is that between dinitrogen and dihydrogen:

$$N_2 + 3H_2 \longrightarrow 2NH_3$$

This reaction is complicated by the participation of four molecules, i.e. one dinitrogen and three of dihydrogen, and necessarily takes place in stages. The world production of ammonia by the Haber-Bosch process, which makes use of this reaction, is around 100 million tonnes per annum. This makes the ammonia molecule number one in the world's synthetic chemicals league table. The reaction is not easily carried out, mainly because of a kinetic restriction. There are four molecules which have to react together to give the two product molecules and a four-body collisional process is most unlikely. The N≡N bond is the strongest triple bond between two identical atoms. The high bond strength and the collisional difficulties combine to make the reaction very slow unless it is carried out at around 400°C in the presence of a finely divided metallic iron catalyst. Because the reaction is one in which four gaseous molecules become two, the entropy change is negative, the order of the system increasing. This is disadvantageous to the reaction and is counteracted by carrying it out at very high pressures of around 200 atmospheres. Even so, the conversion of reactants into products is only 15%. The ammonia produced is removed from the reaction mixture by cooling it, the remainder of the dinitrogen/dihydrogen mixture being recycled. The reaction conditions impose severe engineering problems for the construction of appropriate reaction vessels which are made from special steel. The uses of ammonia are in the manufacture of fertilizers (75%), nitric acid, which is used to make ammonium nitrate fertilizer, the monomer molecules from which various types of nylon are made, and as a refrigerant.

It has been estimated that an even larger quantity of ammonia, 175 million tonnes per annum, is synthesized from atmospheric dinitrogen by various bacteria and blue-green algae, the source of hydrogen being water. The organisms carry out the process at temperatures found in the soil, under normal atmospheric pressure, the partial pressure of dinitrogen being 0.78 atmospheres, and with the aid of an enzyme catalyst called nitrogenase. The bacteria, strains of microorganisms called *Azotobacter*, exist in the roots of growing plants and provide nitrogen, in the form of the ammonium ion, for the manufacture of plant proteins. Efforts are being made to mimic the process in the laboratory, but at the present time the bacteria are far more efficient at carrying out the synthesis of ammonia than any industrial method.

Enzymes consist of protein chains which are coiled up in a highly specific manner. They possess sites within the coil which are receptive to particular molecules and allow the reaction of such molecules to occur much more rapidly than it would in the absence of the enzyme catalyst. Some enzymes are used in domestic clothes washing powders. They operate in the alkaline conditions in a washing machine and are called protein digesters. Some humans develop allergic reactions to the enzymes or to their products in the washed clothes, the remedy being to use non-enzymic detergent powders instead.

Another aspect of catalysis is featured by the catalytic converters which are soon to be legally required to be fitted to all new road vehicles. They consist of a mesh of aluminium alloy with a coating of platinum and rhodium metals. At the high temperatures of the exhaust system, the precious metals catalyse the conversion of nitrogen oxides into dinitrogen by reaction with carbon monoxide:

$$2NO + 2CO \longrightarrow 2CO_2 + N_2$$

$$2NO_2 + 4CO \longrightarrow 4CO_2 + N_2$$

The excess of carbon monoxide is, with the addition of extra air, oxidized to carbon dioxide in a second catalytic stage:

$$2CO + O_2 \longrightarrow 2CO_2$$

Now that lead is no longer added to some petrols/gasolines, the use of catalytic converters makes it possible for vehicle exhaust gases to be a harmless mixture of carbon dioxide and water vapour. It is essential to use lead-free petrol/gasoline because any lead would very quickly poison the catalyst by coating the active metal surfaces.

The daily task of cooking food offers many examples of the applications of reaction kinetics. The most important aspect of reaction kinetics is that which ensures human existence. The human body is an assemblage of organic molecules which are all unstable with respect to carbon dioxide and water, these being the main products of combustion. The conversion of the human body into its combustion products is thermodynamically feasible. That it does not happen spontaneously at normal atmospheric pressure and at the body temperature of 36.9°C is because of the activation energy barrier. The body is maintained by the continual intake of necessary nutrients and gaseous dioxygen. If the maintenance ceases, the transformation of the body into its final stable products is bacterially catalysed unless the provision of heat allows passage over the high thermal barrier to allow the conversion into the final products of the oxides of carbon, hydrogen, nitrogen and sulfur, and a small amount of inorganic ash.

Epilogue

Des and Maggie Carter benefited by reading the book. They understood something about the impact of chemistry upon their lives. They were convinced that the industrial revolution would not have occurred without the prior developments in chemical knowledge. They thought the book showed the extent to which modern society was dependent upon the further development of the subject. They realized that life would be very different if chemistry had not been allowed to have its effect. Without chemistry their existence would be far less acceptable than that they currently enjoyed. They criticized the book for not attempting to measure the quality of life. The quality of life is not to be measured by the amounts of copper dug out of the earth each year. Such activity allows more copper wire to be manufactured and more communications to be made, but surely there must be more in life than the number of television sets. They came to the same conclusions about measuring the quality of life in terms of the numbers of cars, phones, planes and supermarkets, or even the population of the world. Chemistry has had a tremendous effect upon the home comforts which are part of the quality of life, but what about art? There was nothing in the book about this. After considering what they had learned from the book, they decided that art in all its forms would be severely limited without the effect of chemistry.

Painting would be reduced to the use of natural pigments such as were used by the cave painters. There would be no drying oils or synthetic pigments or the popular acrylic paints.

Sculpture would be reduced to natural materials such as wood or stone, the technology of casting metals being precluded. As good as those materials are, the tools with which to alter their shapes would be hard to find.

Music would be sadly reduced to being produced by natural string instruments and drums, there would be no instruments that required metal parts. There would not have been developments in music such as those brought about by Johann Sebastian Bach, Haydn, Mozart, Beethoven, Brahms, Dvorak, Tchaikovsky, Sibelius, Mahler and Shostakovitch, to name just ten. Just imagine what a jazz band would sound like without metal instruments. Steel bands would not exist. There would be no recordings of any performances, the performances would probably not deserve to be remembered in any case.

Theatre would have survived in an outdoor form as there would be no stage lighting and no reasonable make-up. Opera and ballet would not have been developed without chemistry.

Film and cinema would not have happened. Printing would not be possible. This book would not have been produced.

After these deliberations, Des and Maggie were convinced that chemistry had had a very great influence upon their quality of life and were encouraged to study the subject to a higher level.

Appendix

REPRESENTATIONS OF VERY LARGE AND VERY SMALL NUMBERS

Throughout the book, there are quantities which are described either by very large or very small numbers. Printed in their normal fashion, such numbers take up too much space and look unwieldy. To avoid these disadvantages, it is essential to use and to understand the use of scientific notation for numerical quantities.

The largest number that is commonly experienced is the American billion, usually in the context of trade deficits. The billion referred to is that which may be written as 1,000,000,000; one thousand million. Given that most people are accustomed to numbers of this magnitude it is not necessary for scientific notation to be used. When describing a number of pounds sterling such as five billion pounds, it is normal to print the amount as £5 bn or £5 billion. The full version, £5,000,000,000 is rarely used. The reader is thus accustomed to some methods of abbreviating printed quantities. One of the largest numbers encountered in the study of chemistry is that known as Avogadro's number which written out fully is:

602200000000000000000000

or 602,200,000,000,000,000,000,000

taking up a large space. In scientific notation it is written more economically as 6.022×10^{23} meaning that 6.022 is multiplied by 10 twenty-three times. The general form of a number in scientific notation may be written as:

$$n \times 10^{a}$$

where n is a reasonably small number (usually between 1 and 999) and the *power index* or *exponent*, a, indicates the number of times that n should be multiplied by ten to give the number described. In words the number above would be 'n times ten to the power a'. The number 3×10^{8} would be spoken of as 'three times ten to the power eight' or 'three times ten to the eight'. The number five hundred and sixty billion,

560,000,000,000, could be written as 560×10^9 or 56×10^{10} or 5.6×10^{11} or 0.56×10^{12} in scientific notation, all four examples being perfectly acceptable.

Very small numbers are encountered in the study of chemistry, one of the smallest being Plank's constant which is written in scientific notation as 6.626×10^{-34} (in words this is six point six two six times ten to the minus thirty-four). The **negative** index means that the number 6.626 has to be **divided** by 10 thirty-four times to give the Planck constant which in its conventional form is:

$$0.0000000000000000000000000000000006626$$

Just as the large numbers encountered in newspapers describe quantities of pounds sterling or tonnes of wheat or people unemployed, the numbers used in chemistry (and science generally) are descriptions of quantities. The Avogadro number represents the number of molecules of water in eighteen grams of that substance, i.e. one mole of water. A generally agreed system of abbreviating large and small quantities has been developed to avoid writing the 10^a factor in numbers. The prefixes and symbols used to represent the more commonly encountered factors are given in Table A.1.

Table A.1 - Some factors and their prefixes which are used to express very large or very small numbers

factor	prefix	symbol	factor	prefix	symbol
10^3	kilo	k	10^{-3}	milli	m
10^6	mega	M	10^{-6}	micro	μ
10^9	giga	G	10^{-9}	nano	n
10^{12}	tera	T	10^{-12}	pico	p

The prefixes are to be used to quantify units such as those of mass (e.g. grams and tonnes) and length (the metre). One thousand grams, 1000 g, may be written as 1 kg using the prefix k for kilo with the unit of mass (g for gram) to express a mass of one kilogram. Likewise, one million tonnes may be written as 1 Mtonne (one megatonne). The millimetre (mm) is familiar as one thousandth of a metre as is the micrometre (μm) as one millionth of a metre. Although not recommended for general scientific usage, the centimetre (cm) is familiar as one hundredth of a metre.

The multiplication and division of numbers in scientific notation follow rules which are expressed by the following symbolic equations and examples.

$$n \times 10^a \text{ multiplied by } m \times 10^b = n \times m \times 10^{a+b}$$

$$\text{e.g. } 3 \times 10^8 \times 2 \times 10^5 = 6 \times 10^{8+5} = 6 \times 10^{13}$$

$$n \times 10^{a} \text{ divided by } m \times 10^{b} = (n \div m) \times 10^{a-b}$$

$$\text{e.g. } 3 \times 10^{8} \div (2 \times 10^{5}) = (3 \div 2) \times 10^{8-5} = 1.5 \times 10^{3}$$

Readers tempted to try out these rules might do a division such as: $4 \times 10^{9} \div (2 \times 10^{9})$ and arrive at the result: 2×10^{0} and should then realize that the value of 10^{0} (ten to the power zero) is 1.

LOGARITHMS

Logarithms are associated with the scientific notation used for the representation of large and small numbers. The definition of a logarithm is that it is the power (or exponent) to which 10 (for normal decadic logarithms) has to be raised to give the number in question. In symbolic terms this may be expressed by the equation:

$$\text{number} = 10^{\text{logarithm of number}}$$

Because $100 = 10^{2}$, the logarithm of 100 is 2, and because $1000 = 10^{3}$, the logarithm of 1000 is 3. Logarithm Tables are incorporated in most scientific calculators so that the logarithm of any number may be obtained by pressing the appropriate button. Logarithms apply equally well to numbers which are smaller than 1; the logarithm of 10^{-4} is -4, etc. A peculiarity of logarithms is that the logarithm of 1 is zero, i.e. 1 is taken to be equal to 10^{0}. The logarithmic scale is a convenient method of using small numbers to express larger ones. Table A.2 below shows a list of numbers which successively increase by factors of ten. The corresponding logarithms increase by units.

Table A.2 - Some numbers and their logarithms

number	logarithm
0.001	-3
0.01	-2
0.1	-1
1	0
10	1
100	2
1000	3
10000000000	10

The sequence of numbers 1, 2, 3, 4,.... is an example of linear behaviour, the difference between any adjacent numbers being the same (1 in this case). The sequence 1, 10, 100, 1000,... is an example of *exponential* behaviour, each number being ten times the magnitude of the previous one. The logarithms of the numbers of the exponential series (0, 1, 2, 3,....) exhibit linear behaviour. The logarithms are the *exponents* or powers to which ten has to be raised in order to represent each number. Exponential series are not restricted to successive numbers which increase by a factor of ten. Ten is the normal basis of the logarithmic scale, sometimes described as the decadic scale, because of its use in normal arithmetic with the ten integers ranging from 0 to 9. Many other examples of exponential behaviour exist. The natural logarithmic scale, obtained via the 'ln' button on scientific calculators, is based on a multiplier with the value 2.718281828..., this being a number known as *e* by mathematicians. It is useful in chemical theory. A more practical example of exponential behaviour, discussed fully in Chapter 4, is the decrease in atmospheric pressure with altitude, the height above sea level. Because of the influence of gravity upon the atmosphere the pressure decreases by a factor of ten for every 16 km of altitude.

Further reading

HISTORY

The Fontana History of Chemistry, by William H. Brock, Fontana Press, 1992.

CHEMISTRY

GCE 'A' Level

Chemistry in Context, by Graham Hill and John Holman, 3rd Edition, Thomas Nelson & Sons, 1989.

Chemistry and Chemical Reactivity, by John C. Kotz and Keith F. Purcell, 2nd Edition, Saunders College Publishing, 1990.

University level

Introductory

Understanding Inorganic Chemistry, by Jack Barrett, Ellis Horwood, 1990.

Chemistry and the Living Organism, by Molly M. Bloomfield, 5th Edition, John Wiley and Sons Ltd, 1992.

Modern Inorganic Chemistry, by William L. Jolly, 2nd Edition, McGraw-Hill, 1991.

Advanced

Advanced Inorganic Chemistry, by F. Albert Cotton and Geoffrey Wilkinson, 5th Edition, John Wiley and Sons Ltd, 1988.

Organic Chemistry, by Robert Thornton Morrison and Robert Neilson Boyd, 4th Edition, Alleyn and Bacon Inc, 1983.

Physical Chemistry, by P.J. Atkins, 5th Edition, Oxford University Press, 1993.

GENERAL

Environmental Chemistry, by Peter O'Niell, 2nd Edition, Unwin Hyman, 1993.

Atmospheric Pollution, by Derek M. Elsom, 2nd Edition, Blackwell, 1992.

Chemistry in the Marketplace, by Ben Selinger, 4th Edition, Harcourt, Brace and Jovanovich, 1989.

Dealing with Genes, The Language of Heredity, by Paul Berg and Maxine Singer, Blackwell, 1992.

Chemical Evolution, by Stephen F. Mason, Clarendon Press, Oxford, 1992.

The Consumer's Good Chemical Guide, by John Emsley, W.H. Freeman, 1994.

Glossarial index

Index of industries, substances and natural phenomena

Index of data